Janice VanCleave's

Food and
Nutrition
for Every Kid

Other Titles by Janice VanCleave

Science for Every Kid series
Janice VanCleave's Astronomy for Every Kid
Janice VanCleave's Biology for Every Kid
Janice VanCleave's Chemistry for Every Kid
Janice VanCleave's Constellations for Every Kid
Janice VanCleave's Dinosaurs for Every Kid
Janice VanCleave's Earth Science for Every Kid
Janice VanCleave's Ecology for Every Kid
Janice VanCleave's Geography for Every Kid
Janice VanCleave's Geometry for Every Kid
Janice VanCleave's The Human Body for Every Kid
Janice VanCleave's Math for Every Kid
Janice VanCleave's Oceans for Every Kid
Janice VanCleave's Physics for Every Kid

Spectacular Science Projects series
Janice VanCleave's Animals
Janice VanCleave's Earthquakes
Janice VanCleave's Electricity
Janice VanCleave's Gravity
Janice VanCleave's Insects and Spiders
Janice VanCleave's Machines
Janice VanCleave's Magnets
Janice VanCleave's Microscopes and Magnifying Lenses
Janice VanCleave's Molecules
Janice VanCleave's Plants
Janice VanCleave's Rocks and Minerals
Janice VanCleave's Volcanoes
Janice VanCleave's Weather

A+ Project series
A+ Projects in Biology
A+ Projects in Chemistry

Also
Janice VanCleave's 200 Gooey, Slippery, Slimy, Weird, and Fun Experiments
Janice VanCleave's 201 Awesome, Magical, Bizarre, and Incredible Experiments
Janice VanCleave's 202 Oozing, Bubbling, Dripping, and Bouncing Experiments
Janice VanCleave's Guide to the Best Science Fair Projects
Janice VanCleave's Sourcebook

Play and Find Out series
Janice VanCleave's Play and Find Out about Science
Janice VanCleave's Play and Find Out about Nature
Janice VanCleave's Play and Find Out about Math
Janice VanCleave's Play and Find Out about the Human Body

Janice VanCleave's

Food and Nutrition for Every Kid

Easy Activities That Make Learning Science Fun

JOSSEY-BASS
A Wiley Imprint
www.josseybass.com

Published by Jossey-Bass
A Wiley Imprint
989 Market Street, San Francisco, CA 94103-1741   www.josseybass.com

Published simultaneously in Canada.

Illustrations © 1999 by Laurel Aiello.

The publisher and the author have made every reasonable effort to ensure that the experiments and activities in this book are safe when conducted as instructed but assume no responsibility for any damage caused or sustained while performing the experiments or activities in this book. Parents, guardians, and/or teachers should supervise young readers who undertake the experiments and activities in this book.

Limit of Liability/Disclaimer of Warranty: While the publisher and author have used their best efforts in preparing this book, they make no representations or warranties with respect to the accuracy or completeness of the contents of this book and specifically disclaim any implied warranties of merchantability or fitness for a particular purpose. No warranty may be created or extended by sales representatives or written sales materials. The advice and strategies contained herein may not be suitable for your situation. You should consult with a professional where appropriate. Neither the publisher nor author shall be liable for any loss of profit or any other commercial damages, including but not limited to special, incidental, consequential, or other damages.

Readers should be aware that Internet Web sites offered as citations and/or sources for further information may have changed or disappeared between the time this was written and when it is read.

Jossey-Bass books and products are available through most bookstores. To contact Jossey-Bass directly call our Customer Care Department within the U.S. at 800-956-7739, outside the U.S. at 317-572-3986, or fax 317-572-4002.

Jossey-Bass also publishes its books in a variety of electronic formats. Some content that appears in print may not be available in electronic books.

Library of Congress Cataloging-in-Publication Data

VanCleave, Janice Pratt.
    [Food and nutrition for every kid]
    Janice VanCleave's Food and nutrition for every kid : easy activities that make learning science fun.
        p. cm.
    Includes index.
    Summary: Uses problems, experiments, and activities to present information on a variety of topics related to foods and nutrition.
        ISBN 0-471-17666-4 (cloth).—ISBN 0-471-17665-6 (pbk.)
        1. Nutrition—Juvenile literature.   2. Food—Juvenile literature.
    [1. Food.   2. Nutrition.]   I. Title..
    QP141.V36   1999
    612.3—dc21   98-53677
                                                      CIP
                                                      AC

Printed in the United States of America
FIRST EDITION
PB Printing   10 9 8 7

Contents

This book is dedicated to
Becky Rockey,
Virginia Malone,
Jan R. Pel,
and Laura Roberts

Becky is a director of the Texas Science Hotline, sponsored by the University of Texas Health Science Center at San Antonio. She not only was helpful in finding facts about nutrition for this book, but introduced me to Virginia and Jan, who provided technical assistance with nutrition facts. Virginia Malone is a science assessment consultant and Jan is a researcher at the Netherlands Energy Research Foundation (ECN) in Petten. Last but certainly not least is Laura Fields Roberts, a teacher at St. Matthews Elementary School in Louisville, Kentucky. Thanks Laura, Becky, Virginia, and Jan for your invaluable help.

Introduction

This is a basic book about food and nutrition that is designed to teach facts, concepts, and problem-solving strategies. Each section introduces concepts about food and nutrition in a way that makes learning useful and fun.

Food is any animal or plant substance taken in by living things that is used to provide energy, and promote growth and other life-supporting processes. **Nutrition** is the science that deals with the processes by which living things take in and use food.

This book will not provide all the answers about food and nutrition, but it will offer keys to how you can make wise food choices for good health. It will guide you to answering questions such as, What is the difference between vegetables and fruits? Does sugar make you more active? Why are vitamins important? and Will one rotten apple really spoil the whole barrel of apples?

This book presents food and nutrition information in a way that you can easily understand and use. It is designed to teach nutrition concepts in such a way that they can be applied to many similar situations. The problems, experiments, and other activities were selected for their ability to explain concepts with little complexity. One of the main objectives of the book is to present the *fun* of learning about food and nutrition.

How to Use This Book

Read each chapter slowly and follow procedures carefully. You learn best if each chapter is read in order, as there is some buildup of information as the book progresses. The format for each chapter is as follows:

- **What You Need to Know:** Background information and an explanation of terms.

- **Exercises:** Questions to be answered or situations to be solved using the information from What You Need to Know.

- **Activity:** A project to allow you to apply the skill to a problem-solving situation in the real world.

- **Solutions to Exercises:** Step-by-step instructions for solving the Exercises.

All **boldfaced** terms are defined in a Glossary at the end of the book. Be sure to flip back to the Glossary as often as you need to, making each term part of your personal vocabulary.

General Instructions for the Exercises

1. Study each problem carefully by reading it through once or twice before answering.

2. Check your answers in the Solutions to Exercises to evaluate your work.

3. Do the work again if any of your answers are incorrect.

General Instructions for the Activities

1. Read each activity completely before starting.

2. Collect needed supplies. You will have less frustration and more fun if all the necessary materials for the activity are ready before you start. You lose your train of thought when you have to stop and search for supplies.

3. Do not rush through the activity. Follow each step very carefully; never skip steps, and do not add your own. Safety is of the utmost importance, and by reading each activity before starting, then following the instructions exactly, you can feel confident that no unexpected results will occur.

4. Observe. If your results are not the same as those described in the activity, carefully reread the instructions and start over from step 1.

A Note on Nutrition Recommendations   Throughout the book, I have noted general nutritional recommendations from noted **nutritionists** (scientists trained in the science of nutrition). These recommendations are averages and do not necessarily suit every individual. Since each person's nutritional needs vary, please consult a physician before following these recommendations.

1

Gain and Loss

Why Water Is Essential to Life

What You Need to Know

The food you eat contains nutrients. **Nutrients** are the materials in food that your body needs to grow, have energy, and stay healthy. The amount of nutrients you need depends on your size, age, and activity. A baby doesn't need as many nutrients as you do, because it is smaller and less active than you. Grown-ups need more nutrients because they are bigger than you. Boys and girls of the same age generally require about the

NUTRITIONAL NEEDS

Most —

Nutrient Amounts

Least —

Baby   Girl   Boy   Woman   Man

same amount of nutrients. However, men generally need more nutrients than women of the same age because men are usually larger than women.

While the amount of needed nutrients may vary from person to person, the kinds of nutrients are the same. There are six kinds of nutrients. Each kind has a specific health function and all are important. Four of the six nutrients are called **macronutrients** because your body needs them in large quantities. (*Macro* is a prefix meaning large.) The macronutrients are **water, carbohydrates, fats,** and **proteins**. The remaining two nutrients, **vitamins** and **minerals**, are called **micronutrients** because they are needed in smaller quantities by your body. (*Micro* is a prefix meaning small.)

Carbohydrates and fats are the nutrients that provide energy for your body to perform its daily activities. Proteins are needed for growth and repair. Vitamins and minerals are needed to help the body properly use the other nutrients. See chapters 2 through 6 for more information about carbohydrates, fats, proteins, vitamins, and minerals.

Water is the most abundant nutrient in your body. It makes up 50 to 70 percent of your body's weight. Water is in every **cell** (small building blocks of living things) of your body and in the spaces around the cells. It is also in body fluids such as blood, sweat, and tears. **Sweat** is a common name for **perspiration**, salty water released through tiny holes in the skin called **pores**.

Water is necessary for all body functions. These functions include controlling body temperature through sweating, **digestion** (breaking down foods), transporting (carrying) the other nutrients throughout your body, and removing body wastes such as **urine** (liquid waste) and **feces** (solid waste).

In good health, your body's water **input** (taken in) and **output** (taken out or lost) is about **balanced** (equal). About half of the output water is in urine, and the rest is mostly in sweat and the air you breathe out. Most of the output water is replaced by water in the food you **ingest** (eat or drink). But you should also drink five to six glasses of water daily to make sure your body has enough water. The water you drink can come from water in sodas, and fruit drinks, as well as ordinary water.

If lost water is not replaced, your body will suffer from **dehydration** (an excessive loss of water from the body). When dehydration begins, a part of your brain called the **hypothalamus** sends signals that make you feel thirsty. Your water output increases in warm weather and when you are active, because you sweat more. Thus, these conditions would result in a need to ingest more water to prevent dehydration.

Exercises

1. Which figure, A, B, or C, shows the activity that requires the greatest amount of water over a given length of time?

2. Which figure, A or B, represents the input and output of water that would result in dehydration?

A   B

Activity: SWEATY

Purpose   To show how the body loses water through the skin.

Materials   clear plastic bag large enough to cover your foot
rubber band large enough to fit loosely around your ankle
timer

Procedure

NOTE: During hot weather, perform this experiment indoors where it is cooler.

1. Put the plastic bag over your bare foot.

2. Place the rubber band over the bag and around your ankle. *Caution: The rubber band should fit loosely around your ankle and be removed as soon as the experiment is finished.*

3. Observe the bag after 10 minutes.

Results   The bag looks cloudy because of tiny drops of water on its inside surface.

Why?   The water in the sweat released from the pores of your foot **evaporates** (changes from a liquid to a gas) and **condenses** (changes from a gas to a liquid) on the surface of the bag. For the process of **condensation** to occur, the gas must be cooled, which means it loses heat energy. This occurs when the water **vapor** (gas) touches the cool surface of the plastic bag, resulting in the formation of liquid water drops on the bag. For the process of **evaporation** to occur, the liquid must gain heat energy. This occurs by the removal of energy from your skin by the water. So as the water evaporates from your skin, your skin loses heat energy and is cooled. Sweating is one way your body cools itself. The amount of water you lose from sweating is determined by the amount of cooling your body needs.

Warm weather and hard exercise generally cause you to sweat more to keep your body cool. It is not the amount of sweat that makes your body cooler, but the evaporation of the liquid from your skin.

Your entire body sweats, but more sweat is produced on the soles of your feet, palms of your hands, and armpits. Generally, sweat itself is not smelly, but the waste from **bacteria** (microscopic one-celled living organisms found all around us) that live on moist skin can make sweaty feet smelly.

Solutions to Exercises

1. *Think!*

- The amount of water needed by the body increases with an increase in body activity.
- Which figure represents the most body activity over a given length of time?

  Figure C shows the activity requiring the greatest amount of water.

2. *Think!*

- Dehydration occurs when water lost by the body is not replaced.
- Which figure indicates that more water is lost than gained by the body?

  Figure B represents the gain and loss of water that would result in dehydration.

2
Go Power

What Carbohydrates Do

What You Need to Know

Carbohydrates are your body's most important source of energy. **Carbohydrates** are chemicals in food that are made of carbon, hydrogen, and oxygen, and come mainly from plants. There are two groups of carbohydrates, simple and complex.

Simple carbohydrates are sugars, which tend to taste sweet, form crystals, and dissolve in water. Sugars, also called **saccharides**, occur in nature in fruits, some vegetables, maple sap, and honey. There are two main types of saccharides, monosaccharides and disaccharides. **Monosaccharides** are the most basic type of sugar, having only a single sugar **molecule** (smallest particle of a substance that keeps the property of that substance). (*Mono* is a prefix meaning one.) They are also called **simple sugars**. Examples of monosaccharides include **fructose** (fruit sugar), **glucose** (blood sugar), and **galactose** (found in combination with other simple sugars in milk products).

Disaccharides, or double sugars, are made up of two sugar molecules connected together (*Di* is a prefix meaning two.) Sucrose, lactose, and maltose are examples of disaccharides. **Sucrose** (table sugar) is a combination of glucose and fructose. **Lactose** (milk sugar) is a combination of glucose and galactose. **Maltose** (found in sprouting grain) is a combination of two glucose molecules. Generally, simple carbohydrates break

apart more easily during digestion than do complex carbohydrates.

Complex carbohydrates, or **polysaccharides,** are made of many connected saccharide molecules. (*Poly* is a prefix meaning many.) These carbohydrates may be made up of hundreds or thousands of monosaccharides strung together in long complex chains with many side branches. The two main complex carbohydrates in your diet are **starch** and **dietary fiber.** Both come from plants, and both are made up of glucose molecules. Starch is the main storage form of carbohydrates in plants and can be digested into nutrients used by your body. Dietary fiber is a plant carbohydrate that provides structure for plants, but is not a nutrient. Fiber is, however, an important part of your diet. There are two types of dietary fiber, that which is **soluble** (will **dissolve,** or break apart and spread out in a substance) in water, including pectin, and that which is **insoluble** (won't dissolve) in water, including cellulose. **Pectin** helps lower **cholesterol,** which is a type of fat that can clog **blood vessels** (tubes that carry blood through the body). **Cellulose,** or **crude fiber,** absorbs large amounts of water. This water-soaked fiber is needed for proper removal of feces from your body.

Carbohydrates provide you with energy. All digestible carbo-hydrates, simple or complex, are changed into glucose in your body. Glucose is to the body as gasoline is to a car—it is a fuel molecule. Glucose is called blood sugar because it is the sugar carried by the blood to every cell in your body. Recommended sources of carbohydrates are cereals, breads, potatoes, grains, peas, and beans.

In a car, gasoline is **oxidized** (combined with oxygen), pro-ducing new substances and energy. In your cells, glucose is oxidized producing new substances (carbon dioxide and water) and energy. Gasoline remains in the car's storage tank until needed. But if you have more glucose than is needed for ener-gy, it is changed into **glycogen** (a polysaccharide that is the form in which carbohydrates are stored in animals; called ani-mal starch). When your body needs additional energy, the glycogen is changed back to glucose. A chemical change in body cells in which energy is released by the **oxidation** of glu-cose is called **respiration.** (Respiration also includes breath-ing, which brings in oxygen for oxidation.)

If your body has an excess of glucose, and all the glycogen storage sites are filled, the extra glucose is changed into fat (a greasy substance stored in body cells) and you gain weight. (See Chapter 3, "Fat Facts," for more information about fat.) If you stop eating and use up your glycogen, the fat deposits in your body are used as an alternative energy source and you lose weight. How many carbohydrates are enough? Nutritionists generally recommend that about 60 percent of the calories you eat should be carbohydrates, with most of these coming from complex carbohydrates and natural sugars in fruits and vegeta-bles. **Calorie** (Cal) is the unit of measure for food energy.

Exercises

1. Which figure on the next page, A, B, or C, represents a complex carbohydrate?

2. Which figure, A, B, or C, shows foods containing the recommended source of carbohydrates?

Activity: SWEETER?

Purpose To show how ripeness affects the taste of fruits.

Materials unripe banana
glass of water
overripe banana

Procedure

1. Take a bite of the unripe banana. Chew the banana and make note of its sweetness.

2. Drink some water to wash the banana taste from your mouth.

3. Take a bite of the overripe banana. Chew the banana as before, comparing its sweetness to that of the unripe banana.

Results The overripe banana tastes sweeter.

Why?   Although starches are made from chains of sugars, they do not taste sweet. Bananas and most other **fruits** (part of a plant that contains the seeds) grow sweeter as they ripen because the starches in the fruits digest (break down) into sugars. Starchy **vegetables** (any part of a plant that is eaten: roots, stems, leaves, nuts, and fruits and their seeds), however, such as corn and carrots, grow less sweet after harvesting. This is because as these vegetables ripen, other chemicals that do not taste sweet are formed.

Solutions to Exercises

1. *Think!*

- Complex carbohydrates are polysaccharides, meaning many sugars connected together in long complex chains (ones with side branches).

- Which figure represents a long complex chain of sugars connected together?

  Figure C represents a complex carbohydrate.

2. *Think!*

- Carbohydrates are sugars, starches, and fibers.

- All the figures show foods containing carbohydrates.

- Nutritionists recommend that the carbohydrates you eat come mainly from complex carbohydrates (starches and fiber) and natural sugars in fruits and vegetables.

- Figure A has no fruits or vegetables, and figure C has only fruits.

  Figure B shows foods containing the recommended source of carbohydrates.

3

Fat Facts

How Fat Can
Be Good and Bad

What You Need to Know

Fat is most weight watchers' least favorite word. But the truth is that fat is essential for good health. Fat is needed as stored energy, to transport some vitamins, to keep your skin healthy, to **insulate** (to cover with a material that reduces the passage of heat) your body from the cold, and to cushion your body from injury. Fat is the most efficient source of energy. Each gram of fat provides 9 calories of energy as compared to 4 calories for each gram of protein or carbohydrate.

Fats are made of two kinds of chemicals, fatty acids and glycerol. The molecules of **fatty acids** and **glycerol** are made of **atoms** (the smallest units of a molecule) of carbon, hydrogen, and oxygen. Fats are called **triglycerides** because they are a combination of three fatty acid molecules plus one molecule of glycerol. Some fatty acids are **essential,** meaning the body cannot make them so they must be part of your diet. These **essential fatty acids (EFAs)** can be found in foods such as sunflower seeds, walnuts, leafy green vegetables, and corn, canola, safflower, soybean, and sunflower oils.

While too little fat in your diet is not healthy, too much can also be bad for your health. How much is enough? The answer depends on the type of fat. Fats can be put into two categories, saturated triglycerides and unsaturated triglycerides. The carbons in fatty acids of the triglycerides **bond** (connection between atoms) together, forming a chain. If there are single bonds between all the carbons, the fat is said to be a **saturated triglyceride.** At ordinary room temperatures, saturated triglycerides are solid. Saturated triglycerides are more commonly called **saturated fats.** They are considered bad for your body because they cause a number of health problems, from heart disease to cancer. Most animal fats and some plant fats are high in saturated fats, such as butter and coconuts.

Between the carbons of the fatty acids in an **unsaturated triglyceride**, or **unsaturated fat**, there are one or more carbon-to-carbon double bonds (two bonds). A **monounsaturated triglyceride** has at least one fatty acid with only one double bond and a **polyunsaturated triglyceride** has at least one fatty acid with two or more double bonds. The more unsaturated, the more double bonds, and the more likely it will be liquid. In the liquid phase, a triglyceride is called an **oil**. Fats from plants contain a high percent of unsaturated fats. Examples are corn, peanut, and olive oils. The more unsaturated a fat is, the more healthy it is. While the term *polyunsaturated* on a food label does not indicate how unsaturated the fat in the food is, the food is a healthier choice than one with a saturated fat.

The diagram on the next page indicates sources of each type of fat. Note that each food is listed with the fat type it contains the greatest percent of, but it may also have lower percents of the other fat types. Walnuts, for example, are high in polyunsaturated fats, but do contain some monounsaturated and saturated fats. You can determine the different fats in food by checking the nutrition facts on food packages.

Most nutritionists recommend that total fat intake should be no more than 30 percent of a day's calorie intake, and only 10 percent of the day's total fat should be from saturated fat.

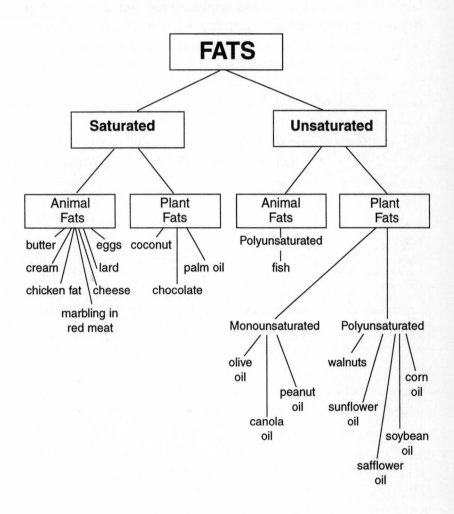

Exercises

1. Which provides more energy, 2 g of fat or 2 g of carbohydrates?

2. If the food you ate in one day contained 2,000 Cal, what
is the greatest number of fat calories you should have
eaten?

3. Study the figure and the nutrition chart to determine who
is eating the candy bar with the fewest harmful fats,
Lauren or Davin.

Candy A

Candy B

Candy A			Candy B		
Nutrition Facts			**Nutrition Facts**		
Serving Size 1 bar			Serving Size 1 bar		
Servings Per Container			Servings Per Container		
Amount Per Serving			**Amount Per Serving**		
Calories	Calories from Fat		**Calories**	Calories from Fat	
		% Daily Value*			% Daily Value*
Total Fat 7.5 g		%	**Total Fat** 7.5 g		%
Saturated Fat 6.0 g		%	Saturated Fat 0.5 g		%
Polyunsaturated Fat 1.0 g		%	Polyunsaturated Fat 6.0 g		%
Monounsaturated Fat 0.5 g		%	Monounsaturated Fat 1.0 g		%
Cholesterol mg		%	**Cholesterol** mg		%
Sodium mg		%	**Sodium** mg		%
Total Carbohydrates g		%	**Total Carbohydrates** g		%
Sugars g			Sugars g		
Other Carbohydrates g		%	Other Carbohydrates g		%
Protein g			**Protein** g		
Vitamin A %	•	Vitamin C %	Vitamin A %	•	Vitamin C %
Calcium %	•	Iron %	Calcium %	•	Iron %
*Percent Daily Values are based on a 2,000 calorie diet.			*Percent Daily Values are based on a 2,000 calorie diet.		

Activity: SEE THROUGH

Purpose To test for fat in food.

Materials desk lamp or other light source
pencil
ruler
brown paper bag
scissors
pen
eyedropper
oil
paper towel
6 food samples: potato chip, carrot, mayonnaise,
  bread, water, apple juice

Procedure

1. Draw eight 2-inch (5-cm) squares on the paper bag. Cut out the paper squares.

2. Prepare 2 test papers: Using the pen, label one square Without Oil and the other With Oil. These papers will be used to determine the amount of fat, from 0 to 100 percent, in the food samples.

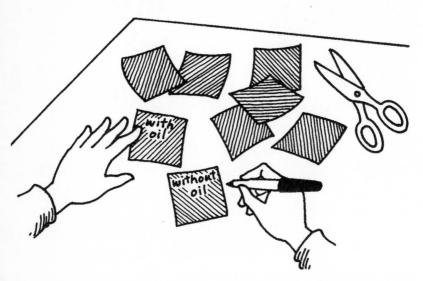

3. Put 1 drop of oil in the center of the paper marked With Oil. Rub the oil across the paper with your finger.

4. Wipe your finger with the paper towel to remove any left-over oil.

5. Label each of the remaining 6 papers with the name of one of the food samples.

6. Rub each solid sample hard across the corresponding paper.

7. Put 1 drop of each liquid food in the center of the corresponding paper. Then, with your finger, rub the food back and forth across the paper.

8. Allow the squares to dry. This should take about 10 minutes.

9. Hold the 2 control papers up to the light. Note the difference in how the light shines through each paper.

10. Compare the sample papers to the test papers: Hold the test papers in one hand and one sample paper at a time in the other hand. Hold the papers up to the light and compare how the light passes through the papers.

Results   Light passing through the papers rubbed with water, apple juice, carrot, and bread matched how light passed through the test paper without oil. Light passing through the papers rubbed with a potato chip and mayonnaise matched how light passed through the test paper with oil.

Why?   In this experiment, the paper without any food rubbed on it showed the known results for 0 percent fat. The paper with oil, a known fat, rubbed on it showed the known results for 100% fat. How light passed through the sample papers was compared to how light passed through the testing papers. Paper becomes more **translucent** (lets some light pass through) when fat is rubbed on it. The fact that fats make paper translucent is used in this experiment to test for the presence of fats in foods.

This test is not always accurate if the amount of fat in a food is very small. For example, bread contains a small amount of fat, but in the author's testing, it did not make the paper translucent. Different types of bread may give different results.

Solutions to Exercises

1. *Think!*

- If 1 gram of fat provides 9 Cal of energy, how many calories will 2 grams of fat provide?

  2 g × 9 Cal per g = 18 Cal

- If 1 gram of carbohydrate provides 4 Cal of energy, how many calories will 2 grams of carbohydrate provide?

  2 g × 4 Cal per g = 8 Cal

  Two grams of fat provides more calories than 2 g of carbohydrate.

2. *Think!*

- No more than 30 percent of daily calories should be from fat.

- Thirty percent of 2,000 calories is how many calories? Note: 30 percent = .30

  0.30 × 2,000 Cal = ?

  For a day's intake of 2,000 Cal, no more than 600 Cal should be from fat.

3. *Think!*

- Both candy bars contain the same amount of total fat.

- Polyunsaturated fat is the best, monounsaturated fat is the second best, and saturated fat is the least healthy type of fat.

- Which candy bar contains the fewest saturated fats?

  Candy bar B contains the fewest saturated fats, so Davin is eating the candy bar with the fewest harmful fats.

4

Linked

The Formation and Importance of Protein in Your Body

What You Need to Know

A large amount of your body, such as skin, nails, hair, blood, and muscles, is made of protein. In fact, every cell in your body contains protein. Proteins are chains of simpler molecules called **amino acids**, containing carbon, hydrogen, oxygen, and nitrogen and sometimes sulfur, phosphorus, and iron. The amino acids are linked together to form long chainlike structures. Every amino acid can link to two other amino acids, one on each end, like boxcars in a long train. There are many different amino acids that can be linked in different order to form thousands of different types of proteins. Some proteins are made of hundreds of amino acids and others are made of thousands of amino acids.

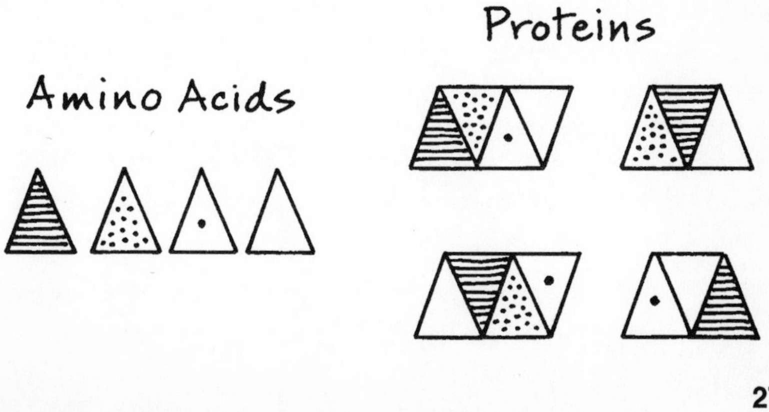

Proteins

Amino Acids

Proteins can be made by both plants and animals. Plants make within themselves all the amino acids and proteins they need. Your body can also make amino acids, but a lot of your amino acids come from protein in the food you eat. This protein is broken down by digestive processes in your body into single amino acids, which are transported to the different body cells. In the cells, the amino acids are combined to form the different proteins needed for growth and cell repair.

Your body needs 20 different amino acids from which to build proteins. Nine of these are **essential amino acids** (amino acids that cannot be made by your body). These must be supplied by the protein in the food you eat.

Proteins in food are one of two types depending on whether they contain all nine of the essential amino acids. A **complete protein** contains all the essential amino acids in the right amounts needed by the body. Sources of complete proteins are poultry, fish, eggs, meat, and **dairy products** (cow's milk and milk products, such as **butter, cheese, cream,** and **yogurt**). An **incomplete protein** lacks some of the essential amino acids. Vegetables (any part of a plant that is eaten) are examples of incomplete proteins. **Legumes** (plants that bear seeds in pods, such as peas, beans, and peanuts) are excellent sources of incomplete protein.

A **vegetarian** is someone who does not eat meat. Some vegetarians, however, do eat animal products such as milk, cheese, eggs, and honey. Vegetables only provide incomplete proteins, but a vegetarian can get the proper amount of essential amino acids by eating **complementary proteins** (combinations of incomplete proteins that provide the nine essential amino acids). Examples of food combinations that provide complementary proteins are cooked dry beans with rice, or peanut butter on whole wheat bread.

Since your body cannot store unused protein or amino acids, you need a daily supply in your diet. Generally for children age 7 to 14, the amount of protein needed is about 0.035 ounces (1 g) daily per 2.2 pounds (1 kg) of body weight. Adults need about 0.028 ounces (0.8 g) daily per 2.2 pounds (1 kg).

Exercises

1. There are nine essential amino acids. Study the figures and select from the following list the combination of two proteins that are a complementary protein.

   a. A and B

   b. A and C

   c. B and C

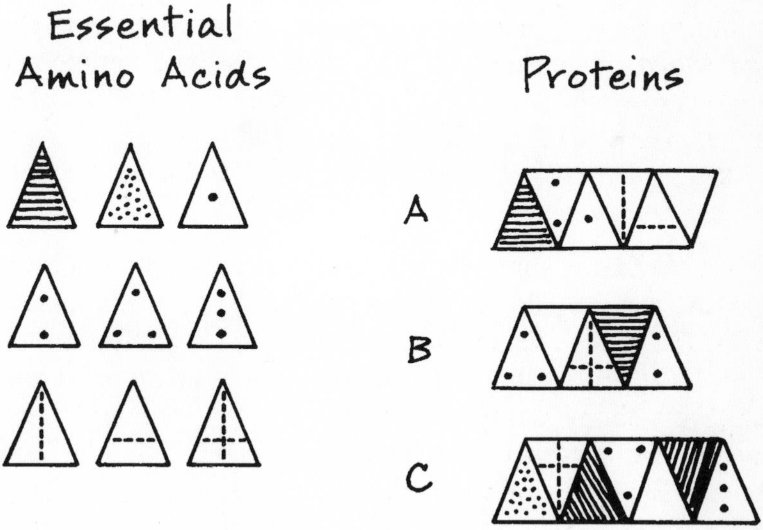

Essential Amino Acids

Proteins

2. Study the figures and determine which one, A, B, or C, represents a source of essential amino acids.

Activity: UNCOILED

Purpose To determine why egg whites become white and foamy when beaten.

Materials 3 egg whites (ask an adult to separate the whites from the yolks)
deep 1-quart (1-liter) bowl
timer
fork
wire whisk (or an electric mixer used with adult assistance)
adult helper

CAUTION: Wash your hands and utensils well when working with raw eggs, as they can contain harmful bacteria.

Procedure

1. Place the egg whites in the bowl and let them stand at room temperature for about 10 minutes.

2. Using the fork, lift the egg whites and observe their appearance.

3. Using the whisk, beat the egg whites until they are stiff. You should be able to make peaks that stand upright on the surface of the mixture by lifting the whisk from the beaten egg whites.

4. Using the fork again, lift the egg whites and observe their appearance.

5. Discard the egg whites.

Results   Before beating, the egg whites are a clear, slightly yellow, thick liquid. After beating, the egg whites are white, thick, and foamy.

Why?   Egg whites are made up of about 87 percent water, 9 percent protein, and 4 percent minerals. (See Chapter 6, "Minerals," for mineral facts.) The kind of proteins in egg whites, called **globular proteins**, are made up of chains of amino acids coiled into compact balls, much like tiny balls of yarn. When the bonds between the amino acids are broken by beating the egg whites, the protein uncoils. This process of changing protein from its natural form is called **denaturing**.

In their natural state, the egg whites are a **viscous** (thick) pale yellow liquid. You see the yellow color because of the way light passes through the clear liquid. Beating the egg whites not only uncoils the protein in them, but forces bubbles of air to be mixed with the uncoiled protein, forming a thick, stiff, white foam. The denatured foamy mass scatters the light in a way that makes the beaten egg whites look white.

Solutions to Exercises

1. *Think!*

- There are nine essential amino acids needed for good health.

- Complementary proteins are combinations of incomplete proteins that provide all nine essential amino acids.

- Which combination provides all nine essential amino acids?

  Choice b represents a complementary protein.

2. **Think!**

   - Complete proteins contain the nine essential amino acids.

   - Sources of complete proteins are poultry, fish, eggs, meat, milk, and milk products.

   - Single vegetables as shown in figures B and C contain incomplete proteins.

   Figure A represents a source of essential amino acids.

5

Alphabet Nutrients

The Importance of Vitamins

What You Need to Know

Vitamins are **organic substances** (substances that contain carbon and come from living things) that your body needs for normal growth and **metabolism** (one or the total of all chemical processes necessary for life). There are many different vitamins and they are named with letters from the alphabet: A, C, D, E, K, and eight different B's (B_1, B_{12}, etc.) Only vitamins D and K can be made in your body. You must eat foods that have the other essential vitamins you need.

Vitamin D helps the body use substances necessary to make bones hard. If you are in the sun for 30 to 60 minutes every week or two, your body can make all the vitamin D it needs. You can also get vitamin D from egg yolks, fish, and fortified foods. **Fortified foods** contain one or more nutrients that do not occur naturally and are added sometime during processing. Milk is fortified with vitamin D.

Vitamin K helps the blood to **clot** (form a mass or lump). About half of the vitamin K your body needs is manufactured by bacteria in your body; the remainder must come from the things you eat. Spinach and other green leafy vegetables, whole grains, potatoes, and cabbage are good vitamin K sources.

Vitamins can be divided into two groups based on the way they are **absorbed** (taken in) into your body. Vitamins A, D, E, and K are **fat-soluble**, which means they dissolve in fat and can be stored in the body. You get them from fatty foods such as oily fish. Vitamin C and the B vitamins are **water-soluble**, meaning they dissolve in water. You get them from watery foods such as fruits. Water-soluble vitamins are not stored by the body for any length of time, but are washed out of the body, mainly in urine. Therefore, foods rich in water-soluble vitamins must be eaten more often than foods with fat-soluble vitamins.

The vitamins you eat are carried by your blood to all of your cells. In the cells, many vitamins combine with enzymes, similar to the way in which puzzle pieces fit together. **Enzymes** are special proteins in living things that control chemical reactions. Some enzymes need a helper vitamin called a **coenzyme** to do their job, which can be to put together or take apart molecules. After the job is finished, the enzyme and vitamin can

be reused many times. This is why large amounts of vitamins are not needed. (See Chapter 19 "Changers," for more information about enzymes and coenzymes.)

Generally, an adequate amount of the essential vitamins is supplied by a healthy diet. But if on occasion you don't eat healthily, some of the needed vitamins can be temporarily supplied by the liver. The **liver** is your body's largest internal organ, weighing 3 to 4 pounds (1.4 to 1.8 kg) in an adult. Part of the liver's job is to store vitamin B_{12} and the fat-soluble vitamins A, D, E, and K. These are released by your body, such as when your diet is lacking in these vitamins.

While many people believe that vitamins are **supermedicinal** (exceptional healing abilities), there is no evidence that this is true. But eating a healthy diet that supplies the necessary vitamins will help keep your immune system in good working order. The **immune system** is the group of body parts that work to make your body resistant to diseases. There is no evidence that for the average healthy American, a daily multivitamin **supplement** (source of nutrients taken in addition to foods) is necessary. However, scientists tend to agree that there is no harm in taking a supplement containing the Recommended Daily Allowance (RDA) of vitamins. (The **Recommended Daily Allowance (RDA)** is the daily amount of the different food nutrients deemed adequate for healthy individuals by the Food and Nutrition Board of the National Research Council, a branch of the National Academy of Sciences.) However, scientists agree that amounts above the RDA can be harmful.

Exercises

1. Which figure on the next page, A, B, or C, shows a good source of vitamin D?

2. Which figure, A or B, represents the way in which many vitamins are used in your body?

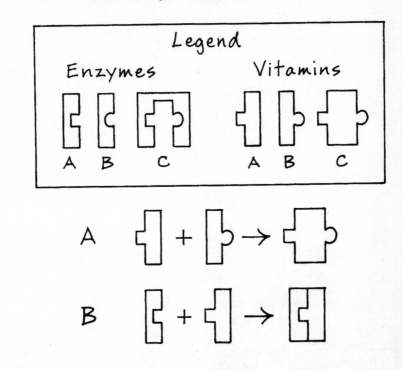

Activity: BROWN BANANA

Purpose To determine if a vitamin can prevent the browning of fruit.

Materials pen
2 paper plates
dinner knife
banana
3 vitamin C tablets (100 mg work well)
cutting board
rolling pin
spoon
timer

Procedure

1. Label one plate With and the other Without.

2. Peel, then slice the banana into 8 pieces. Place 4 slices on each plate.

3. Place the vitamin C tablets on the cutting board and crush them with the rolling pin.

4. Using the spoon, scoop up the crushed vitamins and sprinkle the powder over the cut surface of the banana slices on the plate labeled With.

5. Allow the slices in the Without plate to remain uncovered.

6. Every 30 minutes for 2 or more hours, observe the surface color of each banana slice.

Results   The surfaces of the untreated banana slices slowly turned brown, but the surfaces covered with vitamin C did not change.

Why?   Bananas and other fruits, such as apples and pears, discolor when bruised or peeled and exposed to air. This discoloration is caused by the breaking of cells. The chemicals released by the damaged cells are oxidized, resulting in changes in the fruit that cause it to look brown. Vitamin C is an **antioxidant**, which means that it **inhibits** (decreases or stops) the process of oxidation. Covering the surface of the banana with vitamin C keeps the chemicals in the broken cells of the banana from becoming oxidized, so the fruit doesn't turn brown, or at least it takes a much longer time to turn brown. Since lemons contain Vitamin C, you can add lemon juice to a fruit salad to keep it from browning.

Solutions to Exercises

1. *Think!*

- Good sources of vitamin D are egg yolks, fish, fortified milk, and exposure to sunlight.

Figure B shows a good source of vitamin D.

2. *Think!*

- Figure A shows two vitamins combining to form a third type of vitamin.

- Vitamins do not combine to form new types of vitamins.

- Vitamins combine with enzymes to form a substance that can perform a specific body function, such as breaking apart a molecule that the combined substance would fit into. (See page 141 for another way that coenzymes and enzymes work together.)

Figure B represents the way in which many vitamins are used in your body.

6
Minerals

The Importance of Minerals

What You Need to Know

Minerals are **inorganic substances** (do not contain carbon or come from living things) essential to the functioning of your body. Minerals come from water and soil. In varying amounts, minerals are part of all plants and animals, including humans. Plants take in mineral-rich water from the soil. Animals obtain minerals by eating plants and other mineral-rich animals, and drinking **hard water** (water rich in the minerals calcium, magnesium, and/or iron).

Only small amounts of minerals are needed by your body to be healthy, but some minerals are needed in larger amounts than others. These are called **macrominerals**. The macrominerals are calcium, phosphorus, and magnesium. Minerals needed in smaller amounts are called **trace minerals** or **microminerals**. These include sodium, potassium, chloride, iron, zinc, iodine, copper, manganese, fluoride, chromium, selenium, molybdenum, arsenic, boron, nickel, and silicon.

Minerals are necessary for many life-supporting reactions in your body, such as the function of iron in making **hemoglobin**, the red substance in blood that carries oxygen to cells, and **myoglobin**, the substance that stores oxygen in muscles. The macrominerals, calcium, phosphorus, and magnesium, are responsible for the development of bone, teeth, and muscle **tissue** (a group of similar cells working together to perform a function).

At birth, a baby has about 7 to 9 ounces (200 to 250 g) of calcium in its immature skeleton. As the baby's bones grow, more and more calcium and phosphorus are added, making the bones strong and rigid. By age 20, the bones have generally reached their maximum length and thickness. At this age, the skeletons of both men and women contain about 35 to 42 ounces (1,000 to 1,200 g) of calcium. Most of this calcium is added to the bones during adolescence (age 11 to 18). At about age 50, calcium levels in bones decrease. This can be particularly true for older women who can suffer from a condition known as **osteoporosis**, in which calcium is lost from bones, causing the bones to become easily broken.

Minerals are required for some enzymes to function. Since enzymes are needed for chemical reactions in the body, many chemical reactions would not occur without minerals. Minerals also interreact with each other and with vitamins. For example, copper and vitamin C together help the body absorb iron. Vitamin C helps the body absorb calcium and phosphorus.

When minerals dissolve in water, they break into charged particles called **ions**. The ions formed in body fluids are called **electrolytes**. Electrolytes, such as sodium, potassium, and chloride, help control the movement of water into and out of cells, and from the blood and other tissues to the cells and spaces around the cells. (See Chapter 15, "Salty," for more information about electrolytes.)

Exercises

1. Which figure, A or B, illustrates hard water?

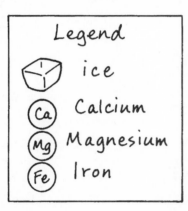

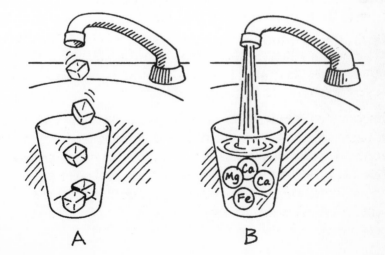

2. Which figure, A, B, or C, represents the age at which bones have the most calcium?

Activity: HARD WATER

Purpose To test for the hardness of water.

Materials measuring spoons
2 tablespoons (30 ml) distilled water
2 baby-food jars with lids
pen
masking tape
¼ teaspoon (1.2 ml) Epsom salts
eyedropper
dishwashing liquid
ruler

Procedure

1. Put 1 tablespoon (15 ml) of water into each of the jars.

2. Use the pen and tape to label one of the jars Soft Water.

3. Add the Epsom salts to the other jar and stir well. Label this jar Hard Water.

4. With the eyedropper, add 3 drops of dishwashing liquid to each jar.

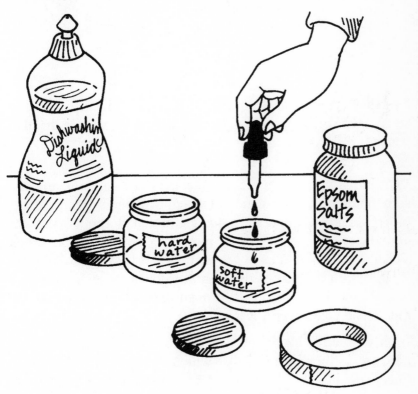

5. Secure the lids on each jar.

6. Holding one jar in each hand, shake the jars vigorously for about 15 seconds.

7. Stand the jars side by side and use the ruler to compare the height of the suds in each.

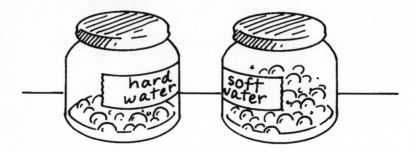

Results   Soapsuds form above the water in each jar. The suds in the soft water are much higher than the suds in the hard water.

Why?   Water that is rich in the minerals calcium, magnesium, and/or iron is called hard water. Calcium is generally the more abundant mineral in hard water, but the greater the amount of any one of these minerals, the harder the water. Water that has few if any of these minerals is called **soft water**. It is difficult to make suds in hard water because the minerals combine with the soap and form **soap scum** (a waxy material that doesn't dissolve in water).

In this experiment, the soft water (distilled water) was made hard by adding Epsom salts, which contains magnesium. In many regions, the drinking water is hard due to minerals in mineral-rich soil that dissolve as rainwater passes through the soil. The quantity of these minerals in drinking water varies in different areas. While minute amounts of minerals that make water hard can be obtained by drinking this water, you obtain more of the minerals by eating plants that take in mineral-rich water from the soil. Drinking water can be a source of the minerals flourine (helps to prevent tooth decay) and copper (helps with iron metabolism).

Solutions to Exercises

1. *Think!*

- The term *hard water* does not mean that the water is actually hard as is ice.

- Hard water is water that has the minerals calcium, magnesium, and/or iron in it.

    Figure B illustrates hard water.

2. *Think!*

- Babies are born with a small amount of calcium in their bones.

- Older adults, particularly women with osteoporosis, have less calcium in their bones than young adults.

- At about age 20, the bones have the greatest calcium content.

    Figure B represents the age at which bones contain the most calcium.

7
Veggies

The Differences between Types of Vegetables, Fruits, Nuts, and Grains

What You Need to Know

Vegetables, fruits, nuts, and grains are foods that come from plants, but they are listed in separate food groups in the Food Guide Pyramid. (See Chapter 8, "Pyramid Power," for information about the Food Guide Pyramid.) A broad definition of vegetable is any edible part of a plant—roots, stems, leaves, fruits, nuts, and seeds. But while fruits, nuts, and seeds can be classed as vegetables, it is not common to do so.

Fruits are the part of the plant that contains the seeds. Fleshy fruits such as apples, berries, and tomatoes have **succulent** (juicy) tissue. Some fleshy fruits, such as tomatoes, squash, cucumbers, and eggplant, are found in the vegetable section of the grocery store.

Other fruits that are commonly called vegetables are legumes. Legumes are fruits in the form of pods splitting along both sides, such as peas, beans, and lentils. A legume that is grouped with nuts because of its nutrient content is the peanut. **Nut** is a popular name for a type of dry, one-seeded fruit of various trees or shrubs having a hard unsplitting shell with a soft kernel inside. Nuts are grouped together, along with peanuts, because of their high amount of protein. Most also have a high amount of fat. Walnuts, pecans, almonds, and Brazil nuts are examples of nuts.

Grains, also called cereals, are the starchy fruits or seeds of various grasses. Common grains are wheat, rice, rye, oats, barley, sorghum, and corn.

Vegetables can be classified according to what part of the plant they come from. Some plants have a single large root, known as a **taproot**. This root not only anchors the plant in the ground, but serves as an underground storage area for food and water for the plant. Edible taproots, called **root vegetables**, include carrots, beets, radishes, yams, sweet potatoes, and turnips.

A **stem** is a plant structure that bears leaves and has **buds** (parts of a stem that develop into other stems or flowers). Stems can grow underground or aboveground. Taro and Idaho potatoes have both underground and aboveground stems. The swollen underground stems are called **tubers**. The tubers store reserve food for the plant.

The aboveground stems of plants support leaves that produce food by **photosynthesis** (a process by which green plants use light energy to change carbon dioxide and water into glucose and oxygen). Examples of aboveground stem vegetables are asparagus and kohlrabi.

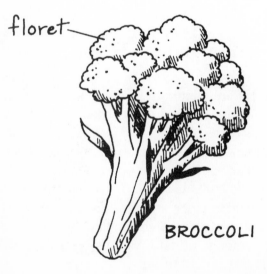

floret

BROCCOLI

Flower vegetables, also called head vegetables, include broccoli and cauliflower. For these two examples, each head is made up of many tiny flowers called **florets**. Generally, these vegetables are eaten while the flowers are in the bud stage. Even though they are not flower vegetables, lettuce and cabbage are commonly referred to as "a head of lettuce" or "a head of cabbage." This is due to their round head shape.

Leaf vegetables include brussels sprouts, cabbage, lettuce, and spinach. The stalks of celery and rhubarb are really just leaf stalks. Bulb vegetables include garlic and onions. Each layer of an onion or garlic bulb is the base of a leaf that stores food and water for the plant. The plant uses up the food stored in the leaves, starting with the outside leaf. As the plant uses the stored food, its fleshy leaf base becomes dry. This is the dry, papery outer skin you see on the outside of onions and garlic.

Exercises

Study the figures to answer the following:

1. Which food is a grain?

2. Which food is commonly called a fruit?

Activity: UP AND DOWN

Purpose To observe the parts of a carrot.

Materials magnifying lens
fresh carrot (with green leaves attached if
available)

Procedure

1. Use the magnifying lens to study the outer surface of the
carrot.

2. Break the carrot in half and observe the cross sections of
the broken halves.

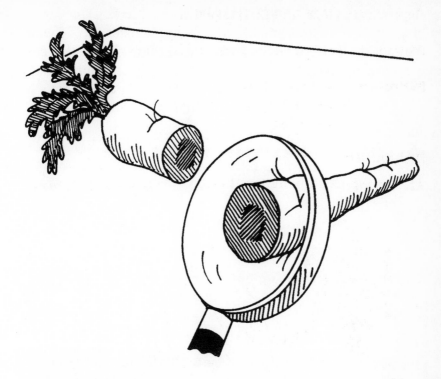

Results   Tiny hairlike fibers are seen on the outer surface of the carrot. The cross section of the carrot reveals two circles, an inner, dark orange circle surrounded by a lighter orange circle.

Why?   The part of the carrot plant you eat is a root. The hairlike structure on the outer surface of this root are the remaining branches of the root system that anchored the plant in the ground.

The inner dark circle in the carrot contains **xylem tubes**, which transport **sap** (water and dissolved nutrients) with dissolved minerals from the soil up into the rest of the plant. The lighter circle around the dark center contains **phloem tubes**, which transport sap with dissolved nutrients, mainly the sugar sucrose, which is basically made in the leaves. The sap in phloem tubes is transported throughout the plant and down to the root. In the root the sucrose is used to make complex carbohydrates, which are stored there. It is the sucrose that makes the carrot taste sweet. The carrot is a taproot in which reserve food and water for the plant are stored.

Solutions to Exercises

1. *Think!*

- Grains are fruits or seeds of various grasses.
- Which of the foods comes from grass? Rice.

  Food A (rice) is a grain.

2. *Think!*

- The sweet, fleshy foods are commonly called fruits.
- Which food is a sweet, fleshy fruit?

  Food D (the pineapple) is commonly called a fruit.

8
Pyramid Power
Guidelines for Daily Food Choices

What You Need to Know

Good nutrition involves following a three-five plan: three key words (moderation, variety, and balance) and five basic food groups. Unless your physician has limited your diet because of a health problem, you can enjoy every food known as long as you practice moderation, variety, and balance. **Moderation** means not extreme. You should eat the correct amounts—not too much and not too little. Too much food results in storage of excess fat as well as excess work done by the body in processing this food. Too little food results in a deficiency of necessary nutrients.

Variety means different kinds. You should eat different kinds of foods. For example, from the fruit group, don't eat just bananas every day. Consider having a banana for breakfast, an apple for a midday snack, and a melon with dinner. Since different foods contain different nutrients, choosing a variety of foods better ensures that you will receive all the nutrients your body requires.

Balance means equal amounts. You need a balance between the number of input and output calories in the foods you eat. To achieve calorie balance, the number of calories that you eat must equal the number of calories your body uses. If input is less than output, then you will lose weight. If input is more than output, then the excess will be stored as body fat and you will gain weight.

The **Food Guide Pyramid** is a diagram that shows the five basic food groups and how much of each food group you should eat each day. The guide shown here was developed by the **U.S. Department of Agriculture (USDA),** a federal agency whose aims include eradicating hunger and malnutrition. The servings on the pyramid indicate a range from the smallest, for a person with an energy need of 1,600 Cal per day, to the largest, for a person with a dietary need of 2,800 Cal per day or more.

FOOD GUIDE PYRAMID

fats, oils and sweets group (use sparingly)

milk, yogurt and cheese group (2-3 servings)

meat, poultry, fish, dried beans, eggs and nuts group (2-3 servings)

vegetable group (3-5 servings)

fruit group (2-4 servings)

bread, cereal, rice and pasta group (6-11 servings)

Based on guidelines developed by the U.S. Department of Agriculture and the U.S. Department of Health and Human Services.

The pyramid shows the number of servings of each food, but what is considered a serving? The Food Group Servings table can be used for school-age children to adults. Smaller servings are needed for younger children.

Food Group Servings

Group	Food	Serving
A	bread	1 slice
	ready-to-eat cereal	1 ounce (28 g)
	cooked cereal, rice, or pasta	½ cup (125 ml)
B	raw leafy vegetables	1 cup (250 ml)
	other cooked or raw vegetables	½ cup (125 ml)
	vegetable juice	¾ cup (188 ml)
C	apple, banana, or orange	1 medium
	chopped, cooked, or canned fruit	½ cup (125 ml)
	fruit juice	¾ cup (188 ml)
D	milk or yogurt	1 cup (250 ml)
	natural cheese	1½ ounces (42 g)
	processed cheese	2 ounces (56 g)
E	cooked lean meat	2 to 3 ounces (56 to 84 g)
	cooked dry beans	½ cup (125 ml)
	egg	1
	peanut butter	2 tablespoons (30 ml)
F	fats, oils, and sweets	use sparingly

The pyramid is designed to indicate which foods should be eaten in the largest amounts. The large base of the pyramid represents the bread, cereal, rice, and pasta group. The servings for this group are more than for any other food group. The top of the pyramid could be called the optional sixth food group. Fats, oils, and sweets are to be eaten sparingly. While small amounts of group F foods are needed, it is not necessary to add them to your diet because most foods naturally contain some of these nutrients.

Remember, your body is like a machine that needs fuel. Eating the right foods provides the fuel necessary to make this wonderful machine work properly, which makes you feel good.

Exercises

1. Which figure, A or B, shows a variety of vegetables?

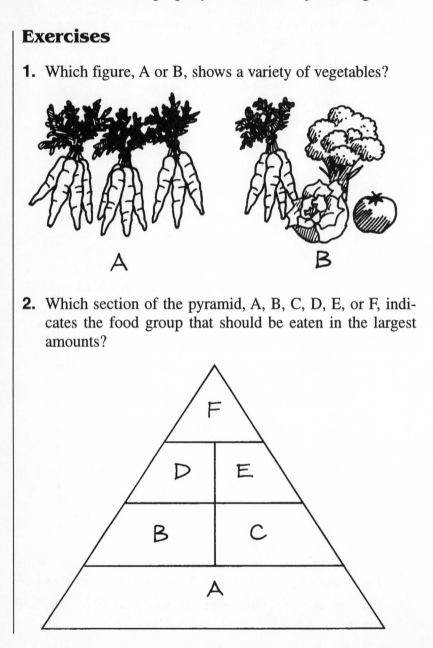

2. Which section of the pyramid, A, B, C, D, E, or F, indicates the food group that should be eaten in the largest amounts?

Activity: THE PYRAMID

Purpose To make a model of the Food Guide Pyramid.

Materials 2 sheets of construction paper, 1 pale color,
  1 white
  ruler
  pen
  scissors
  transparent tape

Procedure

1. Fold the colored paper in half with the short sides together.

2. Unfold the paper. With the ruler and pen, draw two diagonal lines from the fold to the opposite corner as shown to make a pyramid shape.

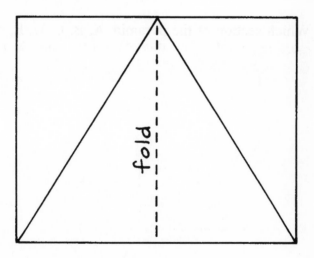

3. Fold the paper in half twice with the long sides together.

4. Cut out the pyramid, then cut along each fold line, to divide the pyramid into four sections.

5. Lay the bottom edge of the pyramid's base (the largest piece of colored paper) along one long side of the white paper. Tape the top edge of the base to the paper.

6. Place the next section of the pyramid above the base and tape the top of this section to the paper.

7. Repeat step 6, taping the remaining colored pyramid pieces to the white paper.

8. Cut the second and third sections along the middle fold line to within ½ inch (1 cm) of the taped edges.

9. Label the paper Food Guide Pyramid.

10. Use the previous Food Guide Pyramid to label the food groups and servings in the appropriate spaces on the pyramid.

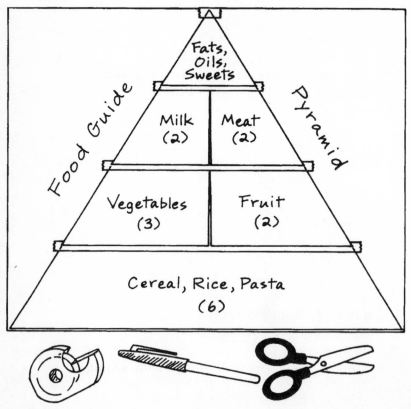

11. Lift the section of paper representing each food group and write the examples of one serving for each group from the following list.

- **Milk**
  1 cup (250 ml) milk or yogurt
  1½ to 2 ounces (42 to 56 g) cheese
  ½ cup (125 ml) ice cream

- **Meat**
  2 to 3 ounces (56 to 84 g) cooked lean meat, poultry, or fish
  1 whole egg
  2 tablespoons (30 ml) peanut butter
  ½ cup (125 ml) cooked dry peas, beans, or lentils

- **Vegetables**
  ¾ cup (188 ml) vegetable juice
  ½ cup (125 ml) cooked or raw vegetables
  1 cup (250 ml) raw leafy vegetables

- **Fruits**
  1 medium apple, banana, or orange
  ½ cup (125 ml) chopped, cooked, or canned fruit
  ¾ cup (188 ml) fruit juice

- **Grains**
  1 slice bread
  1 ounce (28 g) dry cereal
  ½ cup (125 ml) cooked cereal, rice, or pasta

- **Fats, Oils, and Sweets**
  Eat very sparingly.

Results A Food Guide Pyramid with sample servings is made.

Why? The pyramid is a help and a reminder to eat the appropriate servings from the five basic food groups. With enough use, you will memorize the groups and servings. Until that time, keep the pyramid handy when making food selections.

Solutions to Exercises

1. *Think!*

- Variety means different kinds.
- Figure A shows only one kind of vegetable, carrots.

  Figure B shows a variety of vegetables.

2. *Think!*

- The larger the section occupied by a food group on the pyramid, the larger the amount of food that should be eaten.
- The largest section of the pyramid is at the bottom.

  Section A of the pyramid indicates the food group that should be eaten in the largest amounts.

9
Making Choices
Understanding Food Labels

What You Need to Know

Regulations of the **Food and Drug Administration (FDA)** (an agency of the U.S. Department of Health and Human Services that tries to ensure that our foods are safe) require that packaged foods have nutrition labels. These labels must contain specific information, including serving size, total calories and calories from fat, a list of nutrients, and the Percent Daily Value (DV). The **Percent Daily Value (DV)** is the percentage of the suggested daily amount of a nutrient in a food serving based on a 2,000 calorie diet. Over the course of a day, the Percent DVs of nutrients in foods consumed should add up to about 100 percent.

Food labels are intended to show how one serving of the packaged food fits into a daily balanced diet. For example, for a 2,000 calorie diet, nutritionists suggest that only 585 of these calories (65 g) should be fat. The label shown indicates that one serving of this brand of cornflakes contains 0.5 g of fat. To get the Percent DV, divide 0.5 by 65 to get 0.77 percent. On the label, this number is rounded up to 1 percent. For fats, the number of grams is generally more useful than the Percent DV. This is because the number of fat grams for different diets varies.

The food's label indicates a healthy amount of fat. But there are other things to consider before choosing this food. What

Cornflakes A

Nutrition Facts

Serving Size 1 cup (30 g)
Servings Per Container 10

Amount Per Serving

Calories 120   Calories from Fat 5

% Daily Value*

Total Fat 0.5 g	**1%**
Saturated Fat 0 g	**0%**
Cholesterol 0 mg	**0%**
Sodium 290 mg	**13%**
Total Carbohydrate 26 g	**8%**
Sugars 2 g	
Other Carbohydrate 24 g	
Protein 2 g	

Vitamin A 15%	•	Vitamin C 25%
Calcium 4%	•	Iron 45%

*Percent Daily Values are based on a 2,000 calorie diet. Your daily values may be higher or lower depending on your calorie needs:

	Calories:	2,000	2,500
Total Fat	Less than	65 g	80 g
Sat. Fat	Less then	20 g	25 g
Cholesterol	Less than	300 mg	300 mg
Sodium	Less than	2,400 mg	2,400 mg
Total Carbohydrate		300 g	375 g
Dietary Fiber		25 g	30 g

INGREDIENTS: CORNMEAL, SUGAR, WHEAT STARCH, SALT, BROWN SUGAR SYRUP, MALT EXTRACT, CORN SYRUP, CALCIUM CARBONATE

about cholesterol? Your body could not function without cholesterol, but your body can make all the cholesterol that it needs. Foods such as butter, eggs, meat, and shellfish also contain cholesterol. If you eat too much cholesterol, it can end up sticking to the sides of your blood vessels and slowing down blood flow. Many physicians and nutritionists suggest that for a healthy diet, you should eat no more than 300 mg of cholesterol per day. Notice that one serving of the food in the example contains 0 g of cholesterol. This gives the food another positive mark for being a healthy food.

Continue to analyze this food by checking its levels of sodium and sugar. Healthy foods are considered to be those that are low in sodium and sugar. Note the large Percent DV for sodium in the example. Since the chemical name for table salt is sodium chloride, this tells you that there is a lot of salt in this food. The sugar content is low. There is no Percent DV for sugar, but it makes up part of the carbohydrate rating. Of the 29 g carbohydrates, only 2 g is from sugar. The other carbohydrates are starches and dietary fiber. Except for the high sodium content, the food—cornflakes A—is a healthy choice. Before purchasing this cereal, you might look for one with less sodium. (See Chapter 15, "Salty," for information on sodium.)

The ingredients of a food are listed on the label in order from the greatest to the least percentage by weight. This means that for the cornflakes, cornmeal makes up most of the cereal's weight and calcium carbonate the least.

Many packaged foods have labels that say things like "low-fat" or "fat-free." There is now a uniform definition for all of these terms. "Free" means without or with only a very small amount of the nutrient. "Low indicates little, few, or a low source of the nutrient. "Light" means containing one-third fewer calories or one-half the fat of the regular food. "Reduced" or "less" means containing 25 percent fewer calories or nutrients than the regular food. Do beware of "light" products that list fat content by percentage of weight rather than percentage of calories. Percentage of weight can be much less than percentage of calories, making the product appear more healthy than it is.

As you study labels, note the 800 number of the company making the food product. If you do not understand the label, call the 800 number and a representative will gladly answer your questions.

Exercises

Study these labels from two different brands of corn to answer the following questions:

Corn A   **Corn B**

Nutrition Facts		
Serving Size ½ cup (127 g)		
Servings Per Container about 3		
Amount Per Serving		
Calories 100	Calories from Fat 5	
		% Daily Value*
Total Fat 0.5 g		1%
Saturated Fat 0 g		0%
Cholesterol 0 mg		0%
Sodium 430 mg		18%
Total Carbohydrate 22 g		7%
Dietary Fiber 1g		6%
Sugars 11 g		
Protein 2 g		
Vitamin A 2% • Vitamin C 4%		
Calcium 0% • Iron 0%		
*Percent Daily Values are based on a 2,000 calorie diet.		

Nutrition Facts		
Serving Size ½ cup (121 g)		
Servings Per Container about 3½		
Amount Per Serving		
Calories 80	Calories from Fat 5	
		% Daily Value*
Total Fat 0.5 g		1%
Saturated Fat 0 g		0%
Cholesterol 0 mg		0%
Sodium 180 mg		8%
Total Carbohydrate 17 g		6%
Dietary Fiber 2 g		7%
Sugars 4 g		
Protein 2 g		
Vitamin A 0% • Vitamin C 4%		
Calcium 0% • Iron 0%		
*Percent Daily Values are based on a 2,000 calorie diet.		

1. Which corn, A or B, gives you more vitamins?

2. If you are restricting the amount of sodium in your diet, which corn should you eat?

3. Which corn, A or B, contains more sugar per serving?

Activity: CHECK IT OUT!

Purpose To compare nutrition facts of different cereals.

Materials pen
ruler
sheet of typing paper
clipboard

Procedure

1. Use the pen, ruler, and paper to prepare a table similar to the Cereal Nutrition Facts table shown here.

Cereal Nutrition Facts

Brand Name	Calories	Calories from Fat	Sodium	Sugar
1.				
2.				
3.				
4.				
5.				
6.				
7.				
8.				
9.				
10.				

2. Place the table on the clipboard.

3. Take the prepared table and the pen to a grocery store and record nutrition facts for the cereals you usually eat as well as other cereals. Make sure the serving sizes are the same. Add more lines to the Cereal Nutrition Facts table if more than 10 brands of cereal are studied.

4. Study the table and list the cereals in order of calorie content, from least to most. Repeat, making separate lists of the cereals for their nutrient content: fat, sodium, and sugar.

Results   The nutrients of different cereals vary mostly in sodium and sugar content.

Why?   Fat, sodium, and sugar make foods taste better. But for a healthy diet, these nutrients need to be eaten sparingly. The healthiest cereals are low in fat, sodium, and sugar. For weight control, choose a cereal that is low in calories.

Solutions to Exercises

1. *Think!*

- Corn A contains 2 percent vitamin A and 4 percent vitamin C.
- Corn B contains only 4 percent vitamin C.

Corn A contains more vitamins than corn B.

2. *Think!*

- Corn A contains 430 mg of sodium per serving.
- Corn B contains 180 mg of sodium per serving.

Corn B is the better choice for a sodium-restricted diet.

3. *Think!*

- Corn A has 11 g of sugar per serving.
- Corn B has 4 g of sugar per serving.

Corn A has more sugar per serving.

10
Input-Output

How to Maintain a Healthy Body Weight

What You Need to Know

A part of your brain called the hypothalamus causes you to feel hungry when your body needs food. **Hunger** is your body's physical need for food. But what happened the last time you smelled cookies baking, saw a sandwich being made, or tasted your favorite food? Did you have a desire to eat? Even when you are not hungry, these senses—smell, sight, and taste—can trigger your desire for food.

If you wish to prevent weight gain or loss, you must keep a balance of input and output energy. Input energy comes from the food that you eat. Output energy is what you need for basal metabolism and activities. **Basal metabolism** is the amount of energy used by the body while resting and/or **fasting** (not eating) to carry out basic functions, such as breathing, making your heart pump, and growth. A person with a high basal metabolism needs more input energy than a person with a lower basal metabolism. A more active person also uses more energy. So people with a higher basal metabolism and people who are more active need more food to provide the needed energy.

Bodies come in all shapes and sizes. One method to evaluate body weight is to calculate your **body mass index (BMI)**. This is your weight in kilograms divided by the square of your

height in meters. The average BMI for boys is 22 to 24; generally, above 30 is overweight and below 19 is underweight. This means that a boy with a BMI of 22 to 24 is at the 50th percentile for boys his age. Being at the 50th percentile means that half the boys at this age are in this same BMI range. For girls the average BMI is 21 to 23; generally, above 31 is overweight and below 17 is underweight. Although a high BMI usually indicates **obesity** (overweight), there are exceptions. For example, an athletic person can have a high BMI because of an increase in muscle mass rather than mass due to fat.

Physicians also divide people into three basic body types: **ectomorph** (elongated, slender build), **mesomorph** (muscular, medium build), and **endomorph** (rounded body with short arms and legs). Because a muscular person generally has a higher BMI, a person with a mesomorph body type might have a higher BMI than a person of equal height with an ectomorph body type.

Exercises

1. Use the following symbols to complete each equation:

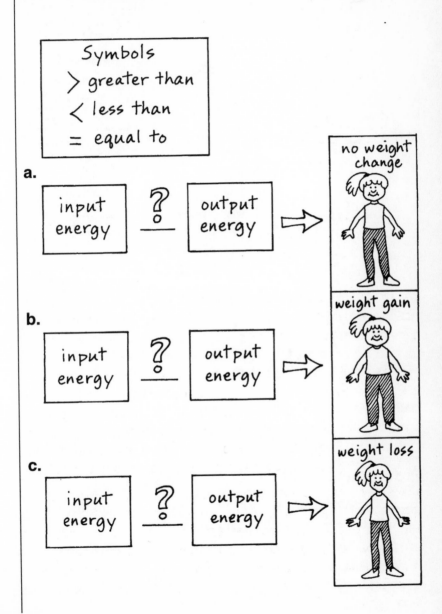

Symbols
> greater than
< less than
= equal to

a.

input energy ? output energy ⟹ no weight change

b.

input energy ? output energy ⟹ weight gain

c.

input energy ? output energy ⟹ weight loss

2. Identify the body types in each figure, A, B, and C, as ectomorph, mesomorph, or endomorph.

A   B   C

Activity: WEIGHTY

Purpose To determine your BMI.

Materials calculator

Procedure

1. Determine your weight in kilograms by dividing your weight to the nearest pound by 2.2.

   Example: If your weight is 88 pounds, your weight in kilograms (kg) would be:

   88 pounds ÷ 2.2 = 40 kg

2. Determine your height in meters by dividing your height in inches by 39.4.

Example: If your height is 53 inches, your height in meters (m) would be:

53 inches ÷ 39.4 = 1.35 m

3. Find your height in square meters (m^2) by multiplying your height in meters by itself.

Example:

1.35 m × 1.35 m = 1.81 m^2

4. Calculate your BMI by dividing your weight in kilograms (step 1) by your height in square meters (step 3).

Example:

40 kg ÷ 1.81 m^2 = 22.09 (BMI)

Results   The BMI calculated from the example weight and height is 22.09.

Why?   With a BMI of 22.09, the example child (boy or girl) falls in the average or 50th percentile. Check with your physician to see if your BMI is average.

Solutions to Exercises

1a. *Think!*

- For no weight change, input energy must equal output energy.
- Use the "equal to" symbol to complete the equation.

  input energy = output energy → no weight change

b. *Think!*

- For weight gain, input energy must be greater than output energy.
- Use the "greater than" symbol to complete the equation.

   input energy > output energy → weight gain

c. *Think!*

- For weight loss, input energy must be less than output energy.
- Use the "less than" symbol to complete the equation.

   input energy < output energy → weight loss

2a. *Think!*

- Figure A shows a body of medium build and muscular.
- Which body type has a medium build and is muscular?

   Figure A is a mesomorph body type.

b. *Think!*

- Figure B shows a rounded body with short arms and legs.
- Which body type has a rounded body and short arms and legs?

   Figure B is an endomorph body type.

c. *Think!*

- Figure C shows an elongated, slender body.
- Which body type is elongated and slender?

   Figure C is an ectomorph body type.

11

Food Changer

How Food Is Digested in Your Body

What You Need to Know

Your **digestive system** is the parts of your body that work together to change the food you eat into particles that are small enough to be taken in by your cells. These particles are the nutrients in food. The changing or breaking down of food by organisms (living things) is called digestion. Digestion can be both mechanical and chemical. **Mechanical digestion** is the physical breaking apart of food into smaller pieces. **Chemical digestion** is the breaking apart of long chains of food molecules into smaller units of combined or separate molecules. For example, glucose, called blood sugar, is one of the important nutrients produced during the digestion of carbohydrates.

Your digestive system is made of a tubular passage through your body called the **alimentary canal** along with various organs that release **digestive juices** (liquids that digest food) into the canal. (**Organs** are groups of tissue that perform the same job.) These organs are the **salivary glands**, located in and near the mouth, and the **pancreas**, liver, and **gallbladder**, all located near the small intestine. The entrance to the alimentary canal is through the mouth. Digestion begins in your mouth, where food is mechanically broken into small pieces by your teeth. Your tongue moves the food around in your mouth, mixing it with saliva released by the salivary glands. **Saliva** is a liquid in the mouth that kills bacteria, and softens,

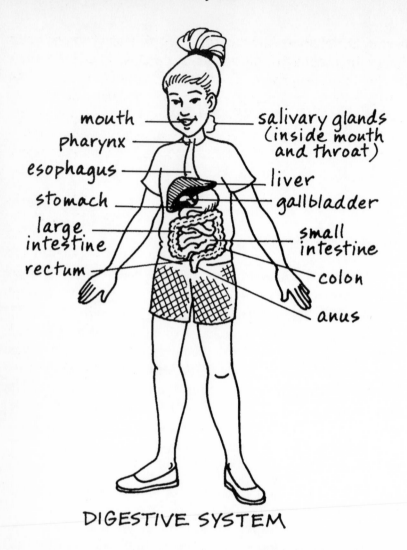

mouth

pharynx

esophagus

stomach

large
intestine

rectum

salivary glands
(inside mouth
and throat)

liver

gallbladder

small
intestine

colon

anus

DIGESTIVE SYSTEM

lubricates, and partially chemically digests the starch in the food you eat. More than 1 quart (1 liter) of saliva is released into your mouth each day.

After the saliva changes the food into a mushy mass, your tongue shapes it into a ball called a **bolus**. When you swallow, your tongue pushes the bolus into your **pharynx** (throat). From here it moves into the **esophagus** (a muscular tube connecting the pharynx and the stomach). To make sure the bolus enters the esophagus, all openings are closed except the one leading to the esophagus. The soft back part of the roof of your mouth, call the **soft palate**, also moves up when you swallow and closes off the **nasal passage** (opening from your nose into your throat). A flap of tissue called the **epiglottis** automatically closes the opening to your **trachea** (the tube through which air passes into the lungs, also called the windpipe).

The bolus does not fall down the esophagus into your stomach. It fits snugly inside the esophagus and is moved forward by the esophagus's muscles. Therefore, you can swallow even if you are upside down (although it is not safe to do so because you can choke on the food). This motion of the esophagus's muscles, called **peristalsis**, is actually waves of muscle **contractions** (a squeezing together) that move food through the digestive system. Peristalsis is much like moving a marble through a rubber tube. Just as squeezing the tube behind the marble moves the marble forward, the squeezing of the esophagus moves the bolus forward.

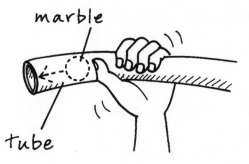

The bolus enters the **stomach** (a pouchlike part of the alimentary canal between the esophagus and the small intestine), where it is **churned** (moved about) and mixed with gastric juices. **Gastric juices** contain an **acid** (a type of chemical) that kills most bacteria in the stomach and small intestine. (See

Chapter 17, "Pucker Up!" for acid facts.) The churned food and gastric acid forms a liquid food mixture called **chyme**. This mixture is pushed into the **small intestine** (part of the alimentary canal between the stomach and the large intestine) a squirt at a time. It takes about 2 to 6 hours after a meal for the stomach to empty into the small intestine. Your small intestine is about 13 to 17 feet (4 to 6 m) long and about 1 to 1½ inches (2.5 to 4 cm) in diameter. This long tube neatly coils around to fit inside your **abdomen** (belly).

Starch is partially digested in your mouth, and proteins are partially digested in your stomach. But most of the food you eat is chemically digested in the small intestine by digestive juices, such as **bile** from the liver and **pancreatic juice** from the pancreas. The large fat molecules enter the small intestine basically unchanged. Here they mix with bile,which is made by your liver and stored in your gallbladder until needed. Just as dishwashing liquid breaks apart grease on dishes, bile breaks fats into tiny globules so enzymes can digest them. Pancreatic juices are able to digest many of the remaining large molecules of carbohydrates, fats, and proteins.

Most of the water and nutrients from the digested food pass through the wall of your small intestine into your **bloodstream** (blood flowing throughout your body). After the nutrients are removed, the remaining part of the chyme is a watery mixture containing mostly undigested plant material (fiber). This mixture enters the **large intestine** (the part of the alimentary canal between the small intestine and the anus). About 99 percent of the water in the chyme entering the large intestine passes into the bloodstream, leaving a semisolid mass of waste called feces.

The large intestine is about 5 to 6 feet (1.5 to 2 m) long and has a diameter about twice that of the small intestine. The large intestine is shaped like an inverted U. The final section of the large intestine, about the last 7 inches (17.5 cm), is called the **rectum**. The end of the rectum that exits the body is called the

anus. The large intestine, except for the rectum, is called the **colon**. After the water is reabsorbed in the colon, the feces is temporarily stored in the rectum until passed from the body through the anus.

Exercises

1. Which figure, A or B, represents the movement of a bolus through the esophagus?

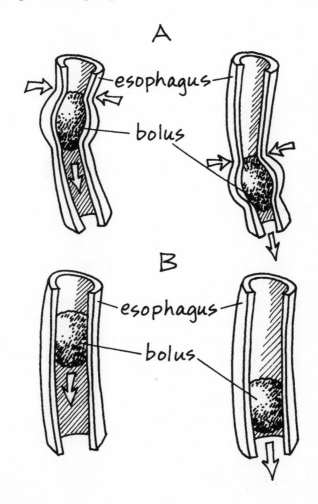

2. Pair each action with the figure that best represents it:

   a. Formation of a bolus

   b. Chemical digestion of food

   c. Mechanical digestion of food

Activity: FOAMY

Purpose To demonstrate how fat is broken into small globules in your body.

Materials desk lamp
pan
2 index cards
2 small cereal bowls
tap water
measuring spoons
2 teaspoons (10 ml) cooking oil
spoon
timer
1 teaspoon (5 ml) dishwashing liquid

Procedure

1. Use the pen to label the cards A, B.

2. Fill the bowls half full with water.

3. Add 1 teaspoon (5 ml) of cooking oil to each bowl.

4. Set the bowls under the desk lamp, each bowl on one of the
   cards. Observe the contents of the bowls.

5. Vigorously stir bowl A with the spoon.

6. Observe the bowl's contents immediately, then again after
   5 minutes.

7. Add the dishwashing liquid to bowl B.

8. Repeat steps 5 and 6 with bowl B.

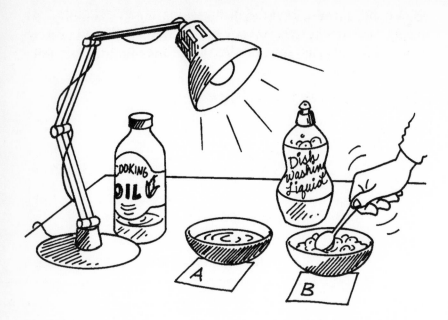

Results Before stirring, the oil generally forms one large thin layer on the surface of the water in each bowl. After stirring, the oil in bowl A broke into small globules and mixed with the water, but after standing the oil separated from the water and formed floating pads of oil on the water's surface. In bowl B, stirring produced a foam and the oil broke into tiny globules that mixed with the water. After standing, some of the tiny globules remained mixed with the water.

Why? Stirring a mixture of two liquids that do not dissolve in each other, such as oil and water, causes one of the liquids (oil) to be suspended (hanging) in little drops throughout the other liquid (water). The result is an **emulsion**. If the liquids separate as in bowl A, the mixture is a temporary emulsion. If an **emulsifier** (substance that prevents an emulsion from separating) such as dishwashing liquid is used, the emulsion does not separate, as in bowl B. In your body, bile acts as an emulsifier. It does the same job on fats in your small intestine that the dishwashing liquid did on the oil in this experiment. Bile breaks the fat into small globules so they can be digested by a fat-digesting enzyme called **lipase**.

Solutions to Exercises

1. *Think!*

- A bolus is moved by peristalsis, the squeezing of muscles in the esophagus.

- Which figure shows the bolus being squeezed through the esophagus?

  Figure A represents the movement of a bolus through the esophagus.

2a. *Think!*

- A bolus is a ball of food shaped by the tongue.

- Which figure shows a ball being formed?

  Figure B represents the formation of a bolus.

b. *Think!*

- Chemical digestion is the breaking apart of long chains of food molecules.

- Which figure shows a chain of connected units breaking apart?

  Figure C represents the chemical digestion of food.

c. *Think!*

- Mechanical digestion is the physical breaking apart of food.

- Which figure shows something being broken into smaller pieces?

Figure A represents the mechanical digestion of food.

12
Tasty

Why Foods Taste Different

What You Need to Know

Stand in front of a mirror and look at your tongue. You will see lots of little red bumps on its surface. These bumps are called **papillae** and they contain many taste buds. **Taste buds** are groups of cells on the tongue and on the roof and the back of

your mouth that are responsible for the sense of taste. Each taste bud is made up of a bundle of cells grouped together like the segments of an orange. The opening at the top of this group of cells is called a **bud pore**. It is through this opening at the top of the taste bud that liquids can enter. Each cell in the taste bud is connected to nerves. **Nerves** are bundles of cells used to send messages to and from the brain and **spinal cord** (bundle of nerves running from the brain down the back). The nerves connected to taste bud cells carry taste messages to your brain.

Another thing to observe about your tongue is that it is wet. This wetness is due to the presence of saliva. Chemicals from food must first dissolve in your mouth's saliva before they can be tasted by the taste buds. You could not taste dry food if your tongue were dry. Determine this for yourself by gently patting your tongue with a paper towel to dry it. Then have a helper sprinkle a few grains of sugar on the dry area. The sugar will have no taste as long as your tongue and the sugar remain dry.

Most of your taste buds are located on the tip, sides, and back of your tongue. Taste buds detect four main tastes—sweet, salt, sour, and bitter. While each taste bud can detect more than one taste, each is best at detecting one of these primary tastes. The taste buds that are best at detecting these tastes are localized in special areas of the tongue.

Sweet is best tasted at the tip of the tongue, salt at the tip and the sides near the tip, sour along the side, and bitter at the back. While there are more of each type of taste bud in specific areas of the tongue, there is considerable overlap of the taste areas and much variation from one person to the next. The flavors of most foods that you eat are a combination of the four tastes of sweet, salt, sour, and bitter.

Much of what we think of as taste is actually smell. Although the messages sent to the brain about the smell and taste of food are independent, your brain blends the messages and voilà! your ice cream not only tastes sweet, but tastes like strawberries. The strawberry **flavor** (taste) is actually due to smell, not to taste buds. The next time you have a cold and your nose is blocked, you will notice that your food is not very tasty.

Your tongue is also sensitive to touch, cold, heat, and pain. Whether the food is wet or dry, soft or hard, smooth or lumpy, hot or cold makes a difference to its taste. Some people like soft apples while others like the hard crunchy ones. Some people like smooth jellies, others like preserves with lumps of fruit. To some people, lumpy mashed potatoes are not as tasty as smooth ones.

Your judgment of how sweet, salty, bitter, or sour a food is also varies with temperature. Tastes are strongest in temperature ranges from about 72° to 105°F (22° to 40°C). Sweet and sour can be judged better at the upper part of this temperature range, and salt and bitter are better judged at the lower part of the range. It is harder to judge the taste of foods that are very hot or very cold.

Exercises

1. Which location, A, B, C, or D, in the figure shows the area of the tongue most sensitive to sweet tastes?

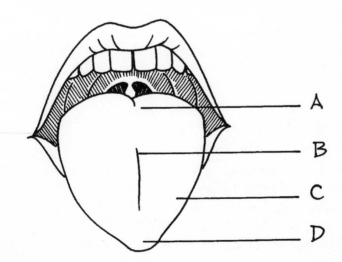

2. Which child, A or B, is able to identify the apple flavor of the apple juice?

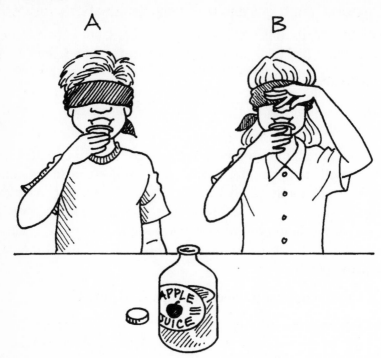

3. Study the figures and choose which food, A or B, when placed on your tongue, will be the first to stimulate the sweet taste buds.

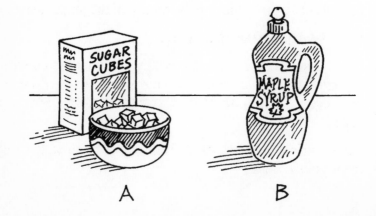

Activity: SWEETER

Purpose   To determine how temperature affects the sweetness of food.

Materials   marker
three 5-ounce (150-ml) paper cups
tap water
measuring spoons
2 tablespoons (30 ml) sugar
spoon
timer

NOTE: This experiment requires a freezer.

Procedure

1. Use the marker to label the cups A, B, and C.

2. Fill cups A and B with cold water. Add 1 tablespoon (15 ml) of sugar to cup B. Stir until all the sugar dissolves.

3. Place cup B in the freezer and leave cup A at room temperature.

4. After 30 minutes, fill cup C with warm water. The water should be warm, but not too hot to drink. Add 1 tablespoon (15 ml) of sugar to the cup. Stir until all the sugar dissolves.

5. Remove cup B from the freezer.

6. Drink a small amount of the cold sugar water from cup B, swishing the liquid around in your mouth so you can taste the sugar. Make note of how sweet the cold sugar water tastes.

7. Take a drink of plain water from cup A to rinse the sugar from your mouth.

8. Repeat steps 6 and 7, using the warm sugar water from cup C.

9. To verify your results, repeat steps 6 through 8 one or more times, changing the order in which you taste the liquids.

Results The warm sugar water in cup C tastes sweeter than the cold sugar water in cup B.

Why? Taste buds that detect sweet tastes are less sensitive to foods at very cold temperatures. Foods that are very cold, such as the cold sugar water in this experiment, must contain more sugar to give them a sweet taste than would the same food served at a warmer temperature. Ice cream and sodas also taste sweeter when warm than chilled.

Solutions to Exercises

1. *Think!*

- Taste buds sensitive to sweet tastes are located on the tip of the tongue.

  Area D in the figure is the area of the tongue most sensitive to sweet tastes.

2. *Think!*

- Identifying a flavor other than salty, sour, bitter, or sweet depends on the smell of the food.

- Because her nose is pinched shut, child B cannot smell the juice.

- Child A can smell the juice.

  Child A can identify the flavor of the apples in apple juice from its taste and smell.

3. *Think!*

- Dry foods cannot be tasted until they are dissolved in saliva and the liquid enters the bud pores.

- The sugar cube would have to dissolve in saliva before it could be tasted.

- The liquid syrup could enter the bud pores and be tasted before mixing with the saliva.

  Food B, maple syrup, would be the first to stimulate the sweet taste buds.

13
Icy

How Ice Affects Foods

What You Need to Know

A glass of water contains millions of tiny particles called molecules, which are made up of atoms. The water molecule is made of two hydrogen (H) atoms connected to one oxygen (O) atom. This is written as H_2O. Water molecules are **polar molecules**, meaning that one side of the molecule has a positive charge and the other side a negative charge. Since an atom has a center with **protons** (positive charges) surrounded by an

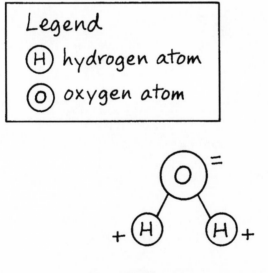

WATER MOLECULE

equal number of **electrons** (negative charges), the atom is **neutral** (no charge). Any loss or gain of electrons upsets this balance. A loss of electrons results in extra positive charges and a gain in a negative charge. Electrons from the hydrogen side of water are pulled closer to the oxygen side. So the hydrogen side is positive and the oxygen side is negative.

Since water molecules are electrically charged, they **attract** (pull toward) other charged particles, including other water molecules. The figure represents the attraction between water molecules. Since unlike charges attract, the positive hydrogen end of one water molecule attracts the negative oxygen end of a second water molecule, forming a bond called a **hydrogen**

LIQUID WATER

bond. The hydrogen of the second water molecule attracts the oxygen of a third water molecule, and so on, forming a group of water molecules. There are usually from four to eight molecules per group in liquid water. The hydrogen bonds between these liquid water molecules are very flexible, allowing the group of molecules to crowd together.

When water freezes, the water molecules combine in an **inflexible** (not bendable) six-sided honeycomb structure. Each structure is a **unit cell**, which is a basic building block of a **crystal** (a solid with its atoms arranged in a definite geometric shape). When ice unit cells join together, they form ice crystals. The more unit cells, the larger the crystal. Small crystals of ice will also join to form larger crystals.

The water molecules in ice crystals are not free to move about and there is more space between the molecules than in liquid water. Because of this, water molecules in ice take up more space than the same number of liquid water molecules.

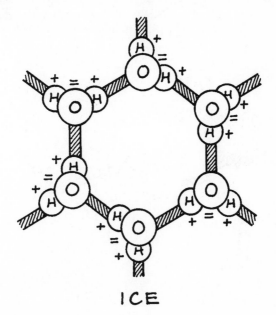

ICE

Foods such as lettuce and apples have large amounts of water in their cells. When these foods are frozen, the expansion (increase in size) of the freezing water within the cells can break the cell wall. When defrosted, the lettuce and apples are no longer crisp.

Other foods, such as ice cream, need tiny crystals of ice. The tiny ice crystals make the ice cream firm, but not hard. The presence of other kinds of molecules in the ice cream, such as those in sugar and milk, help to separate the ice crystals keeping them from joining and growing larger. Other ways of controlling ice crystal size is to continuously stir the mixture while the ice cream is being made. This breaks up the crystals as well as whipping air into the ice cream. The air keeps the crystals separated.

Exercises

1. Figures A and B represent the same food cell. Which figure, A or B, shows the cell when frozen?

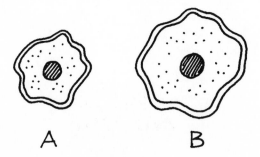

A   B

2. Which figure, A or B, represents ice cream, a frozen mixture of water, milk, sugar, and air?

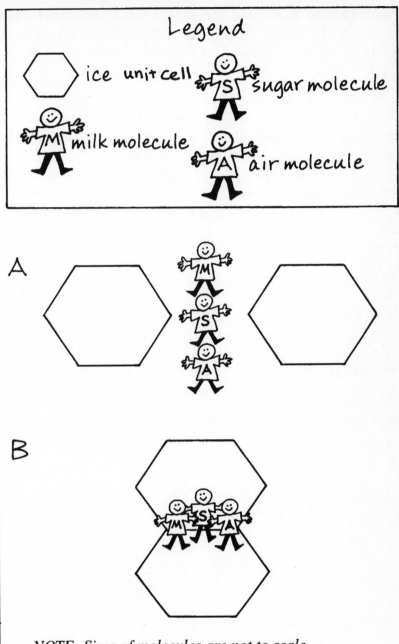

NOTE: Sizes of molecules are not to scale.

Activity: ICY

Purpose To determine why ice pops are softer than ice.

Materials 1-quart (1-liter) jar
tap water
2-quart (2-liter) pitcher
0.15-ounce (4.3-g) package unsweetened flavored
  powdered drink mix
1½ cups (375-ml) granulated sugar
spoon
two 3-ounce (90-ml) paper cups
plate
2 craft sticks

NOTE: This experiment requires a freezer.

Procedure

1. Pour 1 quart (1 liter) of water into the pitcher.

2. Add the powdered drink mix and the sugar. Stir.

3. Place the paper cups on the plate.

4. Fill 1 cup with the flavored drink and the other cup with tap water.

5. Stand 1 craft stick in each cup.

6. Set the plate in the freezer.

7. The next day, remove the plate from the freezer.

8. Remove the paper from the frozen liquids.

9. Holding the craft sticks, carefully try to bite into the ice pop (frozen drink) and the ice (frozen water).

NOTE: Use the remaining flavored drink and 12 more cups and sticks to prepare extra ice pops to eat later or share with friends.

Results The liquid drink and the liquid water both changed to solids, but the frozen drink is not as firm as the frozen water. It is easier to bite into the ice pop than the ice.

Why? The water molecules in each liquid combine to form unit cells that join together to form ice crystals. In the frozen drink, the ice crystals are separated in places by sugar molecules and other ingredients in the drink mix, so the ice crystals in the ice pop are smaller than in the frozen water. These smaller ice crystals make the ice pop softer and easier to eat than the large, solid ice crystal in the frozen water.

Solutions to Exercises

1. *Think!*

- The connections between liquid water molecules are more flexible than those between frozen water molecules.

- Liquid molecules can crowd together and take up less space than can frozen water molecules.

- Cells with frozen water inside expand and can break.

- Which figure shows the cell expanded?

  Figure B represents a frozen cell.

2. *Think!*

- Water freezes, forming unit cells of ice.

- The unit cells are separated by the other kinds of molecules in ice cream.

- Which figure shows unit cells of ice crystals being held apart?

  Figure A represents a frozen mixture of water, milk, sugar, and air.

14
Sweet Stuff

Natural and Artificial Sweeteners

What You Need to Know

A sugar is a carbohydrate that tastes sweet and contains calories. Sugars in naturally occurring sources, such as honey, have been a part of people's diet since prehistoric times. **Refined** (separated and freed from impurities) sugar is a more recent addition to people's diet.

Sucrose is commonly called table sugar or granulated sugar. This sugar was first refined from sugarcane, a giant grass. Sugarcane was the major source of sucrose until the 18th century, when it was discovered that this sugar could be extracted from beets, now called sugar beets. Today, sugarcane and sugar beets are the main sources of sucrose.

Fructose is a sugar found in fruits and honey. It is the sweetest **natural** (found in nature) sweetener. However, the intensity of fructose's sweetness depends on the food it is mixed with. While it may taste sweeter in lemonade, it is not as sweet in cookies or cakes. Most nutritionists agree that there is no nutritional advantage in using fruit juice or honey in place of sucrose. In fact, children less than 1 year old should not be given honey because their immature digestive system may permit the growth of bacteria commonly found in honey.

Corn syrup is another common natural sweetener. This sweetener is made from corn starch, which is treated with enzymes that break the starch into sugars, mostly fructose and glucose. Check the ingredients on different food containers and you will likely find corn syrup listed. This is because corn syrup is so inexpensive to make.

Sugar contains calories and is thus a source of energy. But sugar is said to have "empty calories" because it contains little of the nutrients necessary for good health. If your diet consists of a large amount of candies, cakes, pies, soft drinks, and other sugary foods, you can eat quickly all the daily calories you need without taking in enough of the needed nutrients.

Sugars and other carbohydrates are changed by normal bacteria in the mouth to an acid that can chemically react with **tooth enamel** (hard protective surface) causing it to dissolve. The eating away of tooth enamel is called **dental caries** or tooth decay. The longer this acid is in contact with your teeth, the greater the chance of dental caries. Good dental hygiene such as brushing and flossing after meals has proven to be helpful in reducing the problem. Fluoride (a mineral) in water, toothpaste, mouthwashes, and fluoride treatments applied directly to the teeth increase the resistance of tooth enamel to acid. This is another way to decrease dental caries.

To satisfy the taste for sweets without adding empty calories or promoting tooth decay, many people use **artificial** (man-made) sweeteners. One artificial sweetener is **saccharin**, which is sold under the brand name of Sweet 'n Low. This sweetener is a **synthetic** (not natural; made by the combination of chemicals). While it doesn't add any nutrients to your food, saccharin can be used without adding calories. While saccharin products must have a warning label stating that the product has been determined to cause cancer in laboratory animals, at this time, no human studies have shown an association between saccharin and bladder cancer.

Another calorie-free artificial sweetener is **aspartame**, more commonly known as Equal or NutraSweet. Aspartame is made from the combination of two amino acids, aspartic acid and phenylalanine. Some people fear that artificial sweeteners may be bad for your health in some way. Aspartame is known to affect those who have the disease phenylketonuria (PKU) because they cannot metabolize phenylalanine in the sweetener. Others may be sensitive to the sweetener and need to limit their intake. The FDA has set an acceptable daily intake of aspartame at about 23 mg per pound (51 mg per kg) of body weight.

Exercises

1. Which figure, A, B, or C, shows an artificial sweetener?

A

B

C

2. Which sweetener, A, B, or C, is refined from the plant shown?

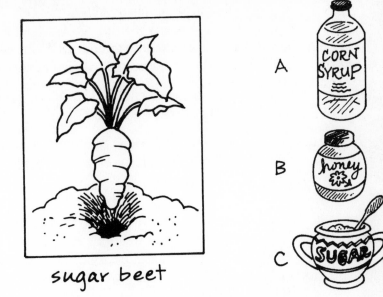

sugar beet

Activity: LESS

Purpose   To compare the sweetness of natural and artificial sweeteners.

Materials   marker
paper plate
packet of NutraSweet
packet of Sweet'n Low
packet (or 1 teaspoon, 5 ml) of table sugar
cup or glass
tap water
3 clean cotton swabs
blindfold
1 resealable plastic sandwich bag
helper

Procedure

CAUTION: Never taste an experiment unless you are sure that it does not contain harmful chemicals or materials.

1. Use the marker to divide the plate into three equal sections. Label the sections NutraSweet, Sweet'n Low, and Sugar.

2. Open the packets of artificial sweetener and sugar and pour them in the corresponding areas of the plate.

3. Fill the cup with water. Then moisten each cotton swab with water and lay one swab on each sweetener in the plate.

4. Place the blindfold over your helper's eyes.

5. Hand one of the cotton swabs to your helper. Instruct your helper to put the swab in his or her mouth, taste the sweetener, and note its sweetness. Place the used swab in the plastic bag.

6. Have your helper drink a small amount of water to rinse the sweetener out of his or her mouth.

7. Repeat steps 5 and 6 with each of the other two sweeteners. Ask your helper to compare the sweetness of each sample. Discard the bag of used cotton swabs.

Results The sweetest-tasting sweetener is Sweet'n Low. Table sugar is the least sweet.

Why? Sweet'n Low is saccharin, a chemical that tastes 300 times sweeter than sucrose. The chemical aspartame, with the trade name NutraSweet, is about 200 times sweeter than sucrose.

Solutions to Exercises

1. *Think!*

   - Artificial sweeteners are synthetic, which means made by the combination of chemicals or not natural.
   - Granulated sugar is a natural sweetener refined from sugarcane or sugar beets.
   - Honey is a natural sweetener produced by bees.
   - Sweet'n Low is chemically produced.

   Figure C shows an artificial sweetener.

2. *Think!*

   - Corn syrup is made from cornstarch.
   - Honey is made by bees.
   - Sucrose is refined from sugar beets.
   - A common name for sucrose is table sugar.

   Sweetener C, table sugar, is refined from sugar beets.

15
Salty

The Function of Sodium in Your Body

What You Need to Know

The term **salt** is commonly used to mean table salt, sodium chloride. When this mineral salt dissolves in body fluids, it breaks apart into two ions, a positively charged sodium ion and a negatively charged chloride ion. Ions in body fluids are called electrolytes. Thus, salt dissolved in body fluids forms the electrolytes sodium and chloride. Chloride is important to the body, but sodium is discussed more often because of its link to **hypertension** (high blood pressure). Sodium does not cause hypertension, but, it does increase the risk for hypertension in people who have inherited a tendency for this disease. If some of your close relatives have high blood pressure, you could have a tendency to have it.

Sodium is absorbed primarily in the small intestine and from there passes into the bloodstream and throughout the body. About 40 percent of your body's sodium is stored in your bones, 50 percent circulates in the blood, and 10 percent is contained in cells. When your blood sodium **concentration** (a measure of how closely packed materials are) is too low, your **kidneys** (organs that filter waste from your blood and produce urine) adjust the amount of urine produced and the amount of sodium in the urine to solve the problem. An increase in the amount of urine causes more water to be removed from your blood. This can result in an increase in the concentration of sodium ions in the blood. A decrease in the amount of sodium in the urine can also result in an increase in the concentration of sodium ions in your blood. When the amount of the sodium in the urine increases, then the concentration of sodium ions in the blood decreases. If the problem cannot be solved by the kidneys, the brain is called upon to tell you to do something. You suddenly feel thirsty if there is too much salt in the blood, or crave a salty snack if more salt is needed.

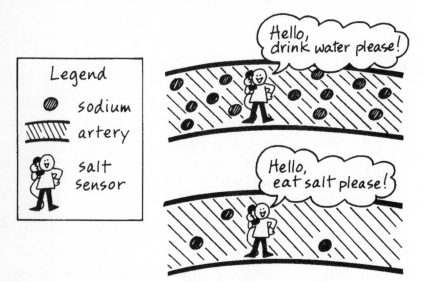

Salt adds the greatest amount of sodium to your diet. It is also found in other foods, such as baking powder. A small amount of sodium is necessary for good health, but the average American school-age child eats about 3,000 to 4,000 mg of sodium per day. While there is no established RDA for sodium, scientists agree that the minimum should be about 500 mg and the maximum about 2,400 mg. (Two-thirds of a teaspoon, or 3.3 ml, of salt supplies the minimum and 1 tablespoon, or 15 ml, supplies the maximum.) While much sodium comes from shaking extra salt onto your food, a lot is also added to food as it is being prepared. You may be surprised to find that many cereals have more sodium than an equal weight of potato chips. This does not mean that you should stop eating cereal.

But it is important to be aware that salt is present in foods that may not taste salty. Read the nutrition labels on packaged foods if you wish to keep count of the salt you are eating.

Foods cooked with salt taste less salty than foods with the same amount of salt sprinkled on them. This is because touching your tongue to the sprinkled salt sends your brain an immediate message that the food is salty. You can satisfy your taste for salt and use less of it by sprinkling salt on food cooked without salt.

Exercises

1. If equal amounts of salt are eaten, which salting method, A or B, produces a food with a saltier taste?

2. Study the figures representing two blood samples to answer the following:

   a. Which figure, A or B, represents the greater concentration of sodium ions?

   b Which figure, A or B, represents the sodium ion concentration that would more likely make you crave a salty snack?

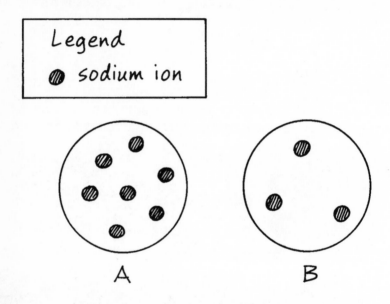

Activity: **WILTING LETTUCE**

Purpose To discover how salt affects water in cells.

Materials masking tape or 2 labels
pen
1 cup (250 ml) tap water
2 small bowls
1 tablespoon (15 ml) table salt
spoon
2 large lettuce leaves
timer

Procedure

1. Use the tape and pen to label the bowls A and B.

2. Pour half of the water into each bowl.

3. Add the salt to the water in bowl A. Stir.

4. Place one lettuce leaf in each bowl.

5. After 30 minutes, pick up the lettuce leaves one at a time and compare their firmness.

Results   The lettuce in the salty water in bowl A feels very limp. The lettuce in water in bowl B is crisp.

Why?   Lettuce is crisp because of water in its cells. Cells in lettuce, like cells in your body, are surrounded by a **selectively permeable membrane**, a barrier that allows some, but not all, materials to pass through. The concentration of salt affects the movement of water into and out of the cell through this membrane. Water moves through the membrane in the direction of lower water concentration. When salt is added to water, the water concentration goes down. Water from inside the lettuce cells moved out of the cells through the membranes into the bowl of salt water, causing the lettuce to **wilt** (become limp).

Sodium ions from salt are an essential part of blood and other body fluids. If you have too much salt in your diet, water can move out of your body's cells into the tissue surrounding the cells. These dehydrated cells do not function properly. But your body is quick to try and fix the problem. One way is to make you feel thirsty and restrict the loss of water from your body. This reduces the amount of saliva in your mouth, making you want to drink even more.

Solutions to Exercises

1. *Think!*

- Food cooked with salt tastes less salty than food sprinkled with salt after it has been cooked.

  Salting method B would produce a saltier tasting food.

2a. *Think!*

- Concentration is a measure of how closely packed materials are.

- The figure showing the greater number of sodium ions in the same amount of blood is the more concentrated.

- Which figure shows more sodium ions?

  Figure A represents a greater concentration of sodium ions.

b. *Think!*

- Craving salt is a response to a low concentration of sodium ions in the blood.

- Which figure has a lesser sodium ion concentration?

  Figure B represents the sodium ion concentration that would cause one to crave a salty snack.

16

Colorful

Natural and Artificial Food Dyes

What You Need to Know

Food manufacturers prepare foods that not only taste good but also look good. To do this they add many of the same natural spices, herbs, and other seasonings, including salt, used in home cooking. But they also add artificial flavors for taste and artificial **dyes** (coloring materials) for appearance.

The FDA has approved about 33 different coloring additives, most of which are natural. However, the natural colorings added to food are more expensive and **unstable** (changeable) than artificial colors. Thus, food manufacturers use mostly artificial colors. Some of the commonly used artificial colors are Red No. 3 (erythrosin), Yellow No. 5 (tartrazine), Citrus Red No. 2, Yellow No. 6, Green No. 3, Blue Nos. 1 and 2, and Red No. 40. Commonly used natural additives and their colors are beets (red), caramel (brown), carotene (orange to yellow), and turmeric (yellow).

While an artificial color may be on the FDA's **GRAS** (Generally Recognized as Safe) list, some may cause problems if consumed in large amounts or by people sensitive to them. For example, tartrazine (Yellow No. 5) is one of the most widely used artificial food colorings. Some people who are extremely sensitive to it break out in **hives** (intensely itchy areas on the skin) and/or have trouble breathing after eating food containing this additive. The FDA requires that tartrazine be identified on the labels of all foods in which it is used. Note that some people may also be sensitive to natural food colorings.

The natural color of foods is due to **pigments** (natural substances that give color). **Anthozanthins** are white pigments found in some vegetables, such as cauliflower, potatoes, rice, and flour. **Carotenes** are vivid yellow and orange pigments that give sweet potatoes, carrots, apricots, and other fruits and vegetables their intense color. It is used to color such foods as butter, margarine, butter-flavored shortening, and cake mixes. **Beta-carotene** is the best-known carotene and not only gives food a bright yellow color, but can be converted by your body into vitamin A. Thus, food labels have "provitamin A" in parentheses after beta-carotene. **Anthocyanins** are red, purple, and blue pigments present in large amounts in some fruits and vegetables, such as raspberries, apples, red and blue plums, red cabbage, and beets.

While one pigment may be the main pigment in a plant, other pigments may also be present. This is true with broccoli, kale, and spinach, which are green due to the presence of the pigment chlorophyll, but also contain carotene. These foods become yellow over time because of the loss of the green chlorophyll that masks the presence of the yellow carotene pigments. (Pigments are lost when they chemically change and no new pigments of the same kind replace them.) The vivid colors of fall leaves are a result of the loss of chlorophyll, which unmasks the other pigments, such as yellow carotenes. Red anthocyanins are produced in the fall.

Exercises

1. Study the coloring ingredients on the box of cereal in the figure. Could this cereal be safely eaten by a person sensitive to tartrazine?

2. Which of the foods shown contain natural coloring additives?

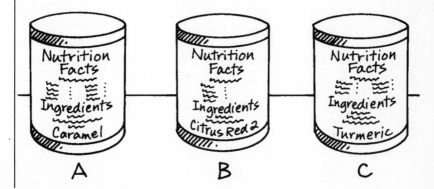

Activity: FOOD PAINTS

Purpose To release pigments from plants to use as fabric dyes.

Materials 6 to 8 sheets of newspaper
white cotton T-shirt
corrugated cardboard (as large as the T-shirt)
pencil
food samples: 4 to 6 spinach leaves,
   2 to 3 beet slices, 4 to 6 blackberries
   or blueberries
waxed paper
lemon-size rock

Procedure

1. Fold the newspaper and place it inside the shirt.

2. Lay the shirt on the cardboard so that the front of the shirt is faceup.

3. Use the pencil to draw a design on the shirt, such as a flower shape with leaves.

4. Lay the spinach over the leaf design, and cover it with a piece of waxed paper.

5. Crush the spinach by tapping it with the rock so that the pigment in the spinach is rubbed into the shirt.

CAUTION: Keep the hand that isn't holding the rock away from the spinach so you do not hit your fingers.

6. Remove the waxed paper and crushed spinach.

7. Repeat steps 4 through 6 to add more color to the same leaf or to color other leaves.

8. Repeat steps 4 through 7, using the other foods to color the flower design.

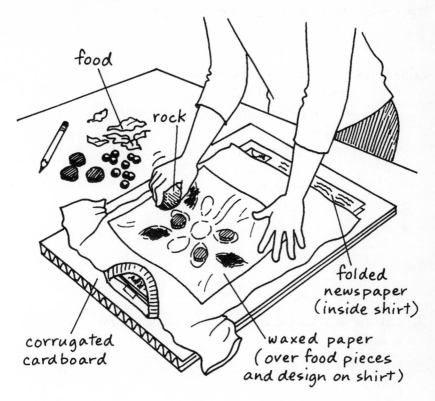

food

rock

folded
newspaper
(inside shirt)

corrugated
cardboard

waxed paper
(over food pieces
and design on shirt)

9. When the design is completely colored, allow the shirt to dry.

NOTE: Washing the shirt by hand in cold water will help keep the colors bright.

Results   Food pigments are used to create a colored design.

Why?   Crushing the food releases the pigments. The pigments get stuck in the fibers of the white cloth and dye the shirt.

Solutions to Exercises

1. *Think!*

- Tartrazine is Yellow No. 5.

- Since tartrazine must be listed as an ingredient on the package of foods containing this food additive, "other color added" does not include tartrazine.

Since tartrazine is not listed as an ingredient, a person sensitive to tartrazine could safely eat the cereal.

2. *Think!*

- The three coloring additives are caramel, Citrus Red 2, and turmeric.

- Which additives are natural? Caramel and turmeric.

Foods A and C contain natural coloring additives.

17

Pucker Up!

The Acids and Bases in Foods

What You Need to Know

Did you know that acids and bases are part of your regular diet? For breakfast you eat a base when you eat the white of an egg, and wash it down with an acid by drinking orange juice. While most of the foods you eat are acidic, there are basic and also neutral (neither acidic nor basic) foods. The special scale for measuring and comparing the amount of acid or base in a food is called the **pH scale**. The values on the pH scale range from 0 to 14, with the pH value of 7 being neutral. Acids are substances with a pH of less than 7 and **bases** are substances with a pH of greater than 7.

Acids and bases are opposites. The mixture of an acid and a base results in a **neutralization** reaction because it produces salt and water, two neutral substances. If a food is acidic, you can reduce or eliminate the amount of acid by adding a base. The reverse is also true.

Acids make foods like citrus fruit, dill pickles, and tomatoes taste sour. The lower the pH, the more sour the taste. The higher the pH, the more basic and the more bitter the taste. Near the neutral pH level, however, acids and bases don't taste especially sour or bitter. Due to the effects of acids or bases, neutral foods have neither a sour nor a bitter taste.

Exercises

Using the pH scale on the next page, answer the following:

1. How many of the following foods—orange juice, egg whites, lemon, tomato, and vinegar—are acidic?

2. Which of the figures A, B, or C represents a neutralization?

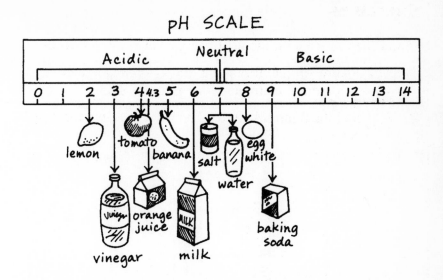

Activity: COLOR CHANGER

Purpose To test for acid in different foods.

Materials 1 cup (250 ml) Anthocyanin Indicator
    (see the Appendix)
1 cup (250 ml) distilled water
1-quart (1-liter) jar
spoon
5 baby-food size jars with lids
masking tape
pen
food samples:
    1 slice each of lemon, tomato
    ½ teaspoon (2.5 ml) each of vinegar,
        baking soda

Procedure

1. Mix the Anthocyanin Indicator and water together in the
large jar, stirring with a spoon.

2. Divide the Anthocyanin Indicator and water mixture among the 5 small jars.

3. Use the tape and pen to label the jars as follows: 1. Lemon; 2. Tomato; 3. Vinegar; 4. Baking soda; 5. Control.

4. Place the corresponding food samples in jars 1 through 4.

5. Close the jars and gently shake them.

6. Compare the color of the liquid in each jar with jar 5, the control.

Results The Anthocyanin Indicator Test Results table shows the colors of the liquid at the start and after addition of the foods.

Anthocyanin Indicator Test Results

Jar	Food Sample	Beginning Color	Color after Adding Food
1	lemon	purple	red
2	tomato	purple	red
3	vinegar	purple	red
4	baking soda	purple	blue-green
5	control	purple	purple

Why? Anthocyanin is a purple pigment found in red cabbage. Other anthocyanins come in colors of red and blue and are found in foods such as berries and plums. Anthocyanin is an **indicator**, which means that its color changes when in the presence of an acid or base. The purple anthocyanin in red cabbage changes to pink or red in the presence of an acid, and to blue or green in the presence of a base. Thus, from the color changes of the Anthocyanin Indicator and water mixture, lemon, tomato, and vinegar are acidic foods and baking soda is basic. The color of the liquid in the control jar was compared to that in the other jars in order to determine the color change.

Solutions to Exercises

1. *Think!*

- Acids have a pH of less than 7.

- What is the pH of each food? Orange juice, pH 4.3; egg whites, pH 8; lemon, pH 2; tomato, pH 4; and vinegar, pH 3.

- How many of the foods have a pH of less than 7?

  Four of the five foods listed are acidic.

2. *Think!*

- Neutralization occurs when an acid and a base are mixed together.

- Which food is an acid? Milk.

- Which food is a base? Baking soda.

- Which figure represents the mixing of milk and baking soda?

  Figure C represents a neutralization.

18
Risers

Leavening and Leavening Agents

What You Need to Know

Leavening is a process by which gases make dough and batter rise. Dough and batter are flour-and-liquid mixtures. **Dough** can be shaped, and **batter** can be poured. There are three types of **leavening gases** (gases that **leaven** (inflate) baked products and make them light and fluffy): carbon dioxide, air, and steam. Carbon dioxide is the most commonly used leavening gas. Substances that produce leavening gases are called **leavening agents**. **Baking soda**, **baking powder**, and **yeast** are leavening agents that when mixed with a liquid, one of the products is carbon dioxide gas.

Baking soda's chemical name is sodium bicarbonate. When this chemical is combined with an acidic liquid, such as **buttermilk**, sour milk, or molasses, carbon dioxide gas is formed.

$$(soda) + (acid + water) \rightarrow CO_2$$

Baking powder is a mixture of baking soda, a dry acid powder, and cornstarch. Double-acting baking powder is the most common leavening agent. Double-acting means that it acts once when moistened and again when heated. The cornstarch is added to absorb water from the air and keep the moisture away from the soda and dry acid. (Rice in a salt shaker does the same thing for salt.) When a liquid, such as water, is added to baking powder, the soda and acid dissolve in the liquid and combine, making carbon dioxide.

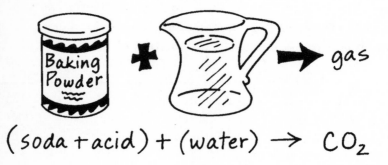

$$(soda + acid) + (water) \rightarrow CO_2$$

The dry acid in double-acting baking powder is generally a blend of the chemicals calcium acid phosphate and sodium aluminum sulfate. These powders react twice with the soda, weakly when mixed with a liquid, then powerfully when heated. The heat not only **activates** (makes more active) the baking powder, but also causes the carbon dioxide bubbles to expand. Baking soda and baking powder are considered quick-acting leavening agents.

Yeast is a slow-acting leavening agent. Yeast is a **fungus**, which is a **microbe** (tiny living thing visible only under a microscope) that slowly digests sugars and starches, producing carbon dioxide gas and **ethanol** (drinking alcohol). Yeast increases the nutritional content of food. It is rich in B vitamins and iron.   One tablespoon (15 ml) of dry yeast contains 1.4 mg of iron, which is about twice that in a large egg.

Yeast works best at temperatures of about 80° to 85°F (27° to 29°C). High temperatures kill yeast and low temperatures slow down its activity. Thus, warm, not hot or cold, liquids are needed to make yeast work best. When yeast dough or batter is baked, heat kills the yeast and expands the bubbles of carbon dioxide gas. The gas bubbles get trapped in small pockets inside the dough or batter. As the gas bubbles expand, the walls of the pockets expand, resulting in the leavening of the mixture. The heat also causes the small amount of ethanol produced by the yeast to evaporate. The nice smell of baking yeast dough is partially due to the evaporation of this alcohol.

Air is also a leavening gas. When egg whites are beaten, air is trapped in the fluffy foam. This foam is carefully added to a flour-and-liquid mixture. Heating causes the air to expand. **Steam** (water in the form of water vapor) can also be used as a leavening gas. Like the other gases, water vapor expands when heated.

Baking leavened batter and dough causes the expanded flour mixtures to harden. Since there are so many gas pockets inside the hardened mass, the bread or other baked product has a spongy look and feel. (See Chapter 20, "Supporter," for an explanation of why bread is soft and spongy.)

Exercises

1. Which choice, A, B, or C, produces a leavening gas?

   A   baking soda

   B   baking soda + water

   C   baking soda + buttermilk

2. Which figure, A, B, or C, shows bread dough with the most inactive living yeast?

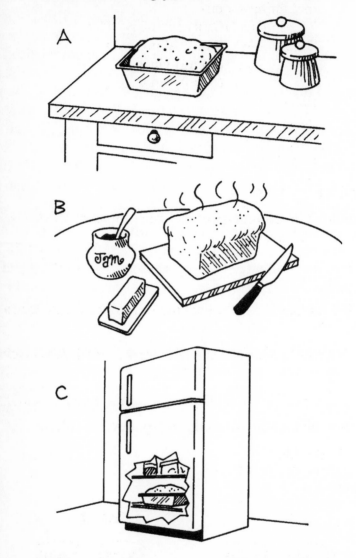

Activity: GASSY

Purpose To study the results of combining the leavening agents baking powder and baking soda with liquids.

Materials marker
eight 3-ounce (90-ml) paper cups
measuring spoons
2 teaspoons (10 ml) baking powder
2 teaspoons (10 ml) baking soda
4 tablespoons (60 ml) vinegar
4 tablespoons (60 ml) tap water
timer
paper
pen or pencil
helper

Procedure

1. Use the marker to number and label the cups BP, BP, BS, BS, V, V, W, and W.

2. Add 1 teaspoon (5 ml) of baking powder to the 2 cups labeled BP.

3. Add 1 teaspoon (5 ml) of baking soda to the 2 cups labeled BS.

4. Pour 2 tablespoons (30 ml) of vinegar in the 2 cups labeled V.

5. Pour 2 tablespoons (30 ml) of water in the 2 cups labeled W.

6. Group the cups into 2 sets of 4 cups each as follows:

   Set 1: BP, BP, V, W
   Set 2: BS, BS, V, W

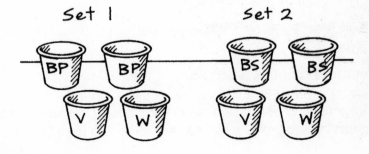

7. Ask your helper to pour the liquids from the cups in set 1 into the cups containing baking powder at the same time that you pour the liquids in set 2 into the cups of baking soda. Pour the vinegar into one cup and the water into the other cup in each set. Do not mix.

8. Observe and record the contents of the cups immediately, then after 30 seconds, and again in 5 minutes.

Results

Test Results of Leavening Agents + Liquids

Leavening Agent + Liquid	Immediately	After 30 Seconds	After 5 Minutes
1. BP + V	much foam	some bubbles	same
2. BP + W	some foam	same	same
3. BS + V	much foam	no bubbles	same
4. BS+ W	no foam	same	same

Why? Vinegar is a weak acid, meaning that it has a small amount of acid mixed with water, which is neutral (neither acidic nor basic). Vinegar reacts very quickly with the soda in the baking powder and baking soda, producing large bubbles of carbon dioxide gas. These bubbles of gas quickly mix with the liquid in each cup, making foam. In a short time, the acid in the vinegar is used up and bubbling slows in the baking-powder-and-vinegar mixture and stops in the baking-soda-and-vinegar mixture. The baking-powder-and-vinegar mixture continues to bubble slowly because of the presence of other acids in baking powder.

Water reacts with the dry acids in the baking powder to produce an acidic solution that slowly reacts with the soda in the baking powder. Bubbles form in the baking-powder-and-water mixture, but not in large amounts as in the baking-powder-and-vinegar mixture. Water is neutral; thus, it does not react with baking soda and no carbon dioxide is produced.

Solutions to Exercises

1. *Think!*

- A leavening agent produces a leavening gas, such as carbon dioxide, air, or steam.
- To produce a leavening gas, baking soda must be mixed with a liquid containing an acid.
- Buttermilk is an acidic liquid.

Choice C shows the production of a leavening gas.

2. *Think!*

- Figure A shows bread dough at room temperature, B at high temperature (baked), and C at a low temperature.
- Yeast is most active at temperatures of about 80° to 85°F (27° to 29°C), which is nearest to room temperature.
- Yeast is killed at high temperatures and becomes inactive at low temperatures.
- Yeast is less active at cool temperatures.

Figure C shows bread dough with the most inactive living yeast.

19

Changers

The Effects of Enzymes on Food

What You Need to Know

Enzymes are proteins found in plants or animals that cause chemical reactions or change the speed of the reactions. There are many different kinds of enzymes. Each enzyme has a specific shape, which allows it to help put together or break apart chemicals. It is believed that enzymes act like a holder into which molecules of chemicals fit like puzzle pieces. Some enzymes receive separate molecules that join and then leave the enzyme together as a single molecule.

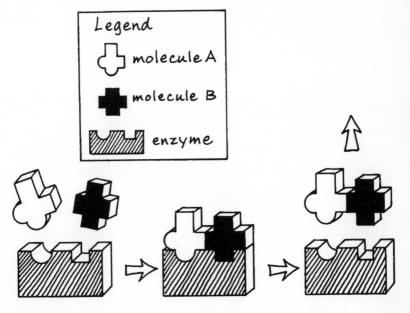

Other enzymes receive a single molecule. These enzymes bend or twist the molecule, causing it to break apart into separate molecules. These separate molecules then leave the enzyme.

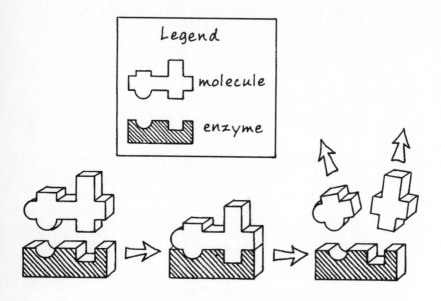

When the separate or single molecules leave an enzyme, the enzyme is then free to receive other molecules. This entire cycle happens so fast that a single enzyme can make about 1,000 changes per second or even faster. Since the enzymes are used over and over and work so fast, a small number of enzymes can cause a lot of chemical changes. Eventually the enzymes wear out and the cells of the organism (any living thing, such as a plant or an animal) make new ones.

Some enzymes need a helper called a coenzyme. The job of some coenzymes is to help an enzyme do its job, which can be to put together or take apart molecules. The figure models the bonding of two molecules, A and B. The coenzymes picks up molecule A and the enzyme picks up molecule B. Because the enzyme and coenzyme fit together, molecules A and B bond, forming the single molecule C. The coenzyme and enzyme separate, molecule C leaves, and the coenzyme and enzyme are free to pick up more molecules. This cycle is repeated again and again until either the enzyme or the coenzyme wears out, at which time the worn part is replaced.

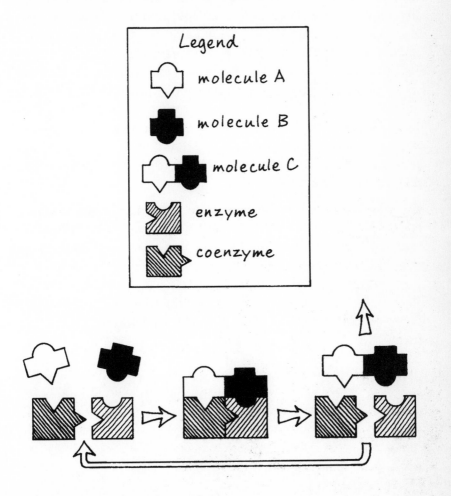

Enzymes cause both desired and undesired changes in food. For example, they are responsible for color, flavor, and texture changes as fruits and vegetables ripen. But they also cause the foods to continue to ripen and eventually rot.

Enzymes can be added to foods to cause changes. **Rennin**, which is an enzyme taken from the lining of a calf's stomach, is used in cheese making. When added to milk, rennin causes the milk protein **casein** to thicken, forming white solid clumps that separate from the liquid. The **coagulated** (thickened) protein is called **curd**. The liquid is called **whey**. The soft curds are dry cottage cheese. This cheese's name may come from the fact that in the past, many farmers made this cheese in their own cottages. Cottage cheese is made from skim milk, which has almost no fat. It is a good low-fat protein food, since 1 cup (250 ml) supplies as much protein as a serving of meat, poultry, or fish. People on low-fat diets should be careful when they purchase cottage cheese, however, because some has cream added, which adds fat.

Enzymes can also be used to make meat more tender. Meats are tough because of the presence of the tough, insoluble protein **collagen**. Meat tenderizers are powders containing enzymes that digest collagen. A common meat tenderizer is **papain** (enzyme extracted from papaya, a tropical fruit). While enzymes digest collagen quickly, there are many molecules of collagen in a piece of meat, and depending on the meat's thickness, it may take 15 minutes or more for the enzymes to work their way through the meat. Extreme temperatures, either hot or cold, **deactivate** (make less active) enzymes. Thus, if cooked soon after adding the enzyme, only a small percentage of the meat's protein will have been digested and the meat will not be very tender.

Exercises

1. In the figures, an equal amount of meat tenderizer is being added to each piece of meat. The clock indicates the time

the tenderizer is added and the time the meat is placed on the grill to cook. Which of the figures, A or B, shows the method that will produce a more tender meat?

2. Observe the figure showing an enzyme cycle. Which symbol, A, B, C, or D, represents the coenzyme?

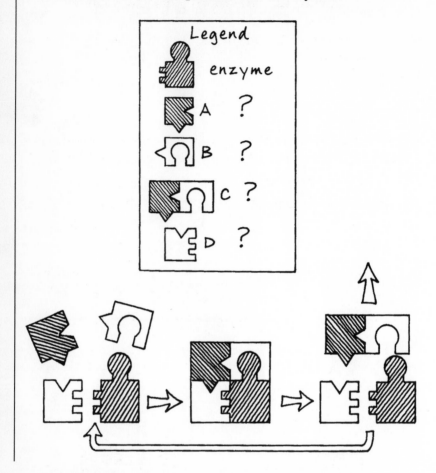

Activity: BREAKDOWN

Purpose To observe the effect of enzymes on gelatin.

Materials 6-ounce (170-g) package of gelatin dessert mix, any flavor
two 1-pint (500-ml) bowls
masking tape

pen
½ cup (125 ml) canned pineapple chunks
spoon
½ cup (125 ml) fresh pineapple chunks
timer
adult helper

Procedure

1. Ask an adult to prepare the gelatin dessert mix according to the instructions on the box, pouring equal portions of the liquid into the 2 bowls.

2. Use the tape and pen to label the bowls 1 and 2.

3. Add the canned pineapple chunks to the gelatin in bowl 1. Stir.

4. Add the fresh pineapple chunks to the gelatin in bowl 2. Stir.

5. Place the bowls in the refrigerator for 3 hours.

6. Remove the bowls from the refrigerator and observe the surface of the gelatin in each bowl. Tilt the bowls slightly and compare their firmness.

Results The gelatin in bowl 1 with the canned pineapple becomes firm, but the gelatin in bowl 2 with the fresh pineapple is less firm or more watery.

Why? **Gelatin** is a gummy protein obtained from animal tissues that is used in making jellylike desserts. Pineapples contain the enzyme **bromelin** which digests the gelatin. If you put raw pineapples in a gelatin dessert, the bromelin in the fruit prevents or inhibits the gelatin's firming. Canned pineapples have been cooked, which deactivates the enzyme. This is why gelatin will get firm if mixed with canned pineapples.

Solutions to Exercises

1. *Think!*

- Collagen is an insoluble protein in meat that causes the meat to be tough.
- Enzymes in meat tenderizer digest the collagen in meat, making the meat tender.
- Heat deactivates the enzymes in meat tenderizer.
- The longer the time before heating, the more meat collagen the enzyme can digest.
- Which figure shows a longer time before heating?

  Figure B shows the method that will produce a more tender meat.

2. *Think!*

- Some coenzymes pick up and carry a molecule to an enzyme, where the coenzyme's molecule attaches to one or more other molecules.

- Which symbol represents a structure picking up a molecule and transporting it to an enzyme?

Symbol D represents a coenzyme.

20
Supporter
Gluten, Flour's Supporting Protein

What You Need to Know

Flour is the main ingredient in most baked foods. It is made from ground grains. A grain is the edible starchy fruit or seed from various grasses, such as wheat, rice, oats, or corn. Most flour is made from wheat grain. A wheat grain is made of **chaff** (the strawlike covering around a grain), **bran** (the nutritious layers forming a protective covering beneath the chaff), **aleurone** (the layer of cells below the bran layers, containing the highest-quality protein), **endosperm** (the starchy interior used to make white flour), and the **germ** (the vitamin- and mineral-rich structure from which a plant develops).

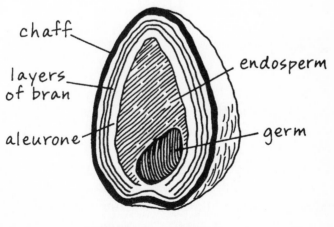

WHEAT GRAIN

Flour is prepared by **milling**, the process of grinding grains and sifting the material to separate the different parts. White flour (also called wheat flour) is made by milling wheat grain to separate the endosperm from the bran and aleurone layers and the germ. This flour is usually **enriched**, which means the important nutrients lost in the milling have been added in amounts about equal to the nutritional content of whole wheat grain. It is usually bleached to make it whiter.

Whole-wheat flour is prepared in the same manner as wheat flour, except that all but the chaff of the grain is ground. This flour is darker, coarser, and more nutritious. Bread made from whole wheat flour is tougher and **denser** (having materials closer together) than white bread. Most whole wheat breads contain a large amount of white flour to make the bread less dense.

There are two types of wheat flour, hard and soft. The hardness of flour is not how hard it feels, but how much proteins it has. Wheat flour that contains a high percentage of protein is called **hard wheat** flour. This is the kind of flour used to make bread. **Soft wheat** flour has less protein and is used to make cakes. Wheats grown in the United States include durum and soft red winter wheat. **Durum wheat** is an especially hard wheat, and flour from this wheat is used for making pasta. Soft red winter wheat is especially low in proteins. Flour made from this wheat is used to produce flaky pastries. The most common flour, made from a mixture of hard and soft wheat, is called **all-purpose flour**.

During the making of dough or batter, a tough, stretchy substance called **gluten** is formed. When liquid is added to flour, the proteins in the flour absorb the liquid, swell, and stick together to form gluten. Mixing the dough or batter stretches the gluten into strands, creating a sticky, elastic mixture. Dough is mixed by **kneading** (pressing, folding, and stretching by hand), and batter is mixed by **beating** (stirring such as with a spoon or electric mixer). Overmixing, which can occur when using an electric mixer, produces too many gluten strands, making the baked product tough.

The gluten strands form a meshlike structure that traps the gases—carbon dioxide, air, or steam—that leaven dough or batter. (See Chapter 18, "Risers," for information on leavening.) The gluten stretches as the heated gases expand, until the heat of baking solidifies the gluten. The baked bread is full of holes where the gases were trapped in the gluten. This is why leavened bread is called "light bread." Observe the surface of a slice of bread to see a solid browned gluten crust and a solid but spongy gluten webbing around gas pockets under the crust.

Exercises

1. Which figure, A or B, shows a mixing method that could result in tough muffins?

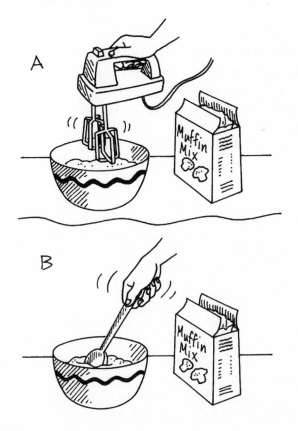

2. Which flour, A or B, is hard flour?

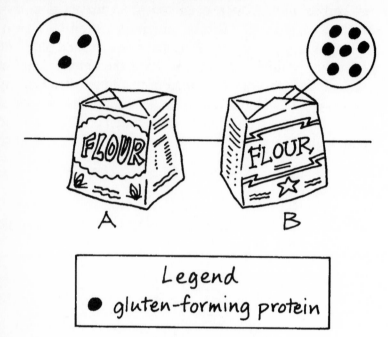

Activity: STRETCHY

Purpose To determine the effect of fat on gluten.

Materials 2 small bowls
spoon
measuring spoons
4 tablespoons (60 ml) all-purpose flour
tap water
fork
1 tablespoon (15 ml) shortening

Procedure

1. In one bowl, mix 2 tablespoons (30 ml) of flour with enough water to make a soft ball. You will need about 2 tablespoons (30 ml) of water.

2. In the other bowl, use the fork to blend together the short-ening and the remaining 2 tablespoons (30 ml) of flour.

3. Add drops of water to the flour-and-shortening mixture until you can make a soft ball.

4. With your hands, shape each ball into a tube-shaped roll about 3 inches (7.5 cm) long.

dough

5. Holding the ends of the flour-and-shortening roll, pull out-ward to stretch it. Observe the ease with which the roll stretches or breaks.

6. Repeat step 5 with the flour roll.

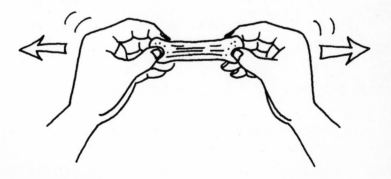

Results The flour roll was stretchier than the flour-and-shortening roll, which broke easily.

Why? Mixing flour and water produces gluten, which gives dough strength and elasticity. Adding fat such as shortening to dough not only improves the flavor and color of the baked product, but also makes it tender by coating the gluten strands so that they don't hold together as tightly. Fat shortens the gluten strands. That's why the term **shortening** is used for the fat often used in making dough.

Solutions to Exercises

1. *Think!*

- Mixing increases the production of gluten. Overmixing can produce too much gluten, making the muffins tough.

- Electric mixers are more likely to overmix than beating by hand.

Figure A shows a mixing method that could result in tough muffins.

2. *Think!*

- Hard flour contains more gluten-forming proteins.

- Which sack of flour has more gluten-forming proteins?

Flour B is hard flour.

21

Easy Chewing

The Changes in Supportive Structures in Foods

What You Need to Know

Placing fruits together speeds up their ripening process because fruits produce **ethylene gas**, a plant hormone that encourages fruit ripening. (**Hormones** are messenger chemicals that are made in one part of a plant or animal and move to another part via plant and body fluids, where they cause a specific response in cells and tissues.) The riper the fruit, the more ethylene gas produced. The saying "One rotten apple can spoil the whole bunch" is true, because the ethylene gas from the overripe apple will speedily cause the other apples to ripen. The discovery that ethylene gas encourages fruit ripening has made it possible for growers to ship unripe fruit to market.

In order to slow down ripening during shipping, fruits are transported in ventilated crates, which have openings to allow gases to flow through. Cool temperatures retard (slow down) the production of ethylene gas. Synthetic ethylene is sometimes used for speedy last-minute ripening. This process is harmless and can even increase the vitamin C content of the fruit. But some people do not like the firmer texture of the synthetically ripened fruit.

Unripened fruits are usually green and firm. As the fruit ripens, not only does the color change, but the firmness of the fruit decreases. The color change comes from the breakdown of the green chlorophyll pigment, which unmasks other pigments in the fruit. The softening of the fruit is due to the breakdown of pectin. Pectin is a water-soluble complex carbohydrate found in a layer between the plant cell walls of plant cells that touch each other. It helps to bind the cells together. In fruits, it helps to make the fruit firm and to keep its shape. When fruits are cooked in water, the pectin dissolves in the hot water, and the cells become less tightly bound together. This also happens when pectin changes to more soluble sugars as the fruit ripens. With age, the continued breakdown of pectin causes the fruit to become sweeter and softer. Eventually the fruit becomes overripe, turning mushy and not very tasty. (Pectin is used commercially in the preparation of jellies and jams. The main sources of commercial pectin are the peels of citrus fruits and apple pulp residue from apple cider presses.)

Just as pectin binds plants cells together, **connective tissue** binds together and supports the tissues of your body. One of

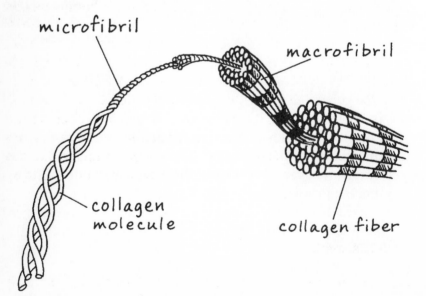

COLLAGEN

the most abundant types of connective tissue is made of fibers, including **collagen fiber**. This fiber is made of collagen molecules, which makes up one-third or more of all your body's protein. A molecule of collagen is made of three protein chains loosely wound together and held by hydrogen bonds. A bundle of many collagen molecules forms a very small fiber called a **microfibril**. A bundle of many microfibrils forms a larger fiber called a **macrofibril**. A bundle of many macrofibrils forms a collagen fiber, which is very strong and very resistant to stretching.

When collagen is boiled, many of the bonds between the protein chains break and gelatin (a protein) forms. The reason tough meats become tender when cooked in liquid is that the insoluble collagen is changed to soft, water-soluble gelatin.

Gelatin is used in cooking to turn a liquid into a semisolid. This is done by adding dry gelatin to a cold liquid containing water. The gelatin absorbs water and swells. Next, the gelatin and the liquid are heated, causing the swollen gelatin to break into particles that spread evenly and become a **homogeneous mixture**, meaning it has the same composition throughout. This mixture is an example of a **colloid** (a homogeneous mixture of particles suspended in a gas, liquid, or solid). If the colloid is composed of solid particles suspended in a liquid, as with the gelatin and water, the colloid is called a **sol**. When the sol is cooled, the gelatin particles link together trapping the liquid in pockets. This process, called **gelling**, turns the sol into a semisolid colloid called a **gel**. The trapped liquid makes the gel wiggly. Heating the gel melts the gelatin and turns the mixture back into a sol.

Exercises

1. In which figure on the next page, A or B, will the fruit ripen faster?

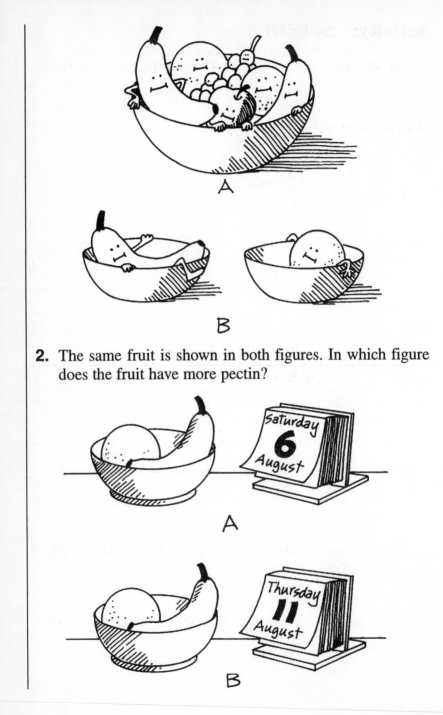

2. The same fruit is shown in both figures. In which figure does the fruit have more pectin?

Activity: SPEEDY

Purpose To discover a method of speeding up the ripening of a banana.

Materials pencil
ruler
paper
2 unripe bananas
paper lunch bag

Procedure

1. Use the pencil, ruler, and paper to make a Ripening of Bananas table for recording your data, such as the one shown here.

Ripening of Bananas

Day	Covered Banana	Uncovered Banana
1		
2		
3		
4		

2. Observe and record the color of both bananas on day 1 (starting day).

3. Place one banana in the bag.

4. Close the bag and place it on a table or kitchen counter.

5. Lay the other banana next to the bag.

6. Observe and record the color of the bananas each day at about the same time for 3 more days. Note how much the appearance of each banana changes—for example, "mostly green with yellow areas" or "about half yellow."

Results   The covered banana in the bag changed from green to yellow more quickly than did the uncovered banana.

Why?   Bananas change from green to yellow as they ripen because they lose their green chlorophyll. Bananas and other fruits produce ethylene gas, which speeds up the ripening process. As the fruit ripens, it produces even more ethylene gas. The ethylene gas produced by the banana in the bag was trapped, causing the banana to ripen even faster. Most of the gas produced by the uncovered banana **diffused** (spread out in all directions) in the air around it before it could affect the banana. Other fruits can be ripened using this method.

Solutions to Exercises

1. *Think!*

- Ethylene gas is a hormone that stimulates ripening of fruit.

- Fruit produces more ethylene gas as it ripens.

- Placing fruit together speeds up the ripening process because there is more ethylene gas present.

  The fruit in figure A will ripen faster.

2. Think!

- Pectin in fruit breaks down and changes to sugar as the fruit ripens.

- Fruit ripens as time passes.

- Which figure shows the younger fruit?

  The fruit in figure A has more pectin.

22
Liquid Nutrient

Why Milk Is an Important Food

What You Need to Know

Milk may not be the most perfect food, but it is certainly one of the best foods for promoting and maintaining health. This is true because it contains so many different nutrients that are needed by people of all ages. It is an excellent source of proteins, vitamins, and minerals, especially calcium.

Milk is produced by all **mammals** (animals that produce milk to nurse their young). Milk from each mammal can be different. In cow's milk there is about an equal balance of fat, proteins, and carbohydrates. Human breast milk is lower in fat than cow's milk, and higher in sugar.

For the first few months, breast milk is the ideal food for infants. Not only does it supply nutrition, but it also provides **antibodies** (disease-fighting proteins) until the baby's body can make enough of these microbe fighters. (Microbes, or microorganisms, are tiny living things visible only under a microscope, such as bacteria. Some microbes can cause disease.) As the baby grows, it needs food other than milk to supply nutrients not in breast milk. A growing child requires a large amount of calcium to build bones and teeth. While milk from one type of mammal may not always be usable by another type, cow's milk can be used by most people. Dairy products are the best sources of calcium for a growing child or a person of any age.

Milk contains the sugar lactose. Before your body can use this sugar, it must be digested into simpler sugars called galactose and glucose that can be used by your body. **Lactase**, an enzyme in your small intestine, is what breaks down lactose. Normally, people have plenty of lactase, but some people's bodies cannot make any lactase and others cannot make enough. People without lactase or too little lactase are said to have a **lactose intolerance**. This disorder is more common among blacks and people of Asian origin.

With age, the bodies of more and more people of all races produce inadequate amounts of lactase. Without lactase to digest lactose, the milk sugar stays in your small intestine, where bacteria causes it to ferment. **Fermentation** is a chemical reaction in which microbes growing in the absence of air cause changes in foods. Gas is one of the products of fermentation. Depending on the microbe, alcohol or an acid may also be produced. The fermentation of lactose changes it into gas and lactic acid. The gas **bloats** (swells up) the intestines and the acid irritates them, resulting in a bellyache and diarrhea.

One solution for lactose intolerance is not to eat milk or most other dairy products. This means no ice cream, ever! But there is good news. The lactose in some dairy products, such as yogurt and aged cheeses, has been predigested by bacteria. You can buy lactose-free milk to drink. But some people prefer to take lactase in tablet form. Even if you are lactose intolerant, you can eat ice cream if you take a lactase tablet. But don't take lactase pills just because eating ice cream gave you a bellyache. Let your physician decide if you have a lactose intolerance, and if so, he or she can advise you on how much lactase is best for you to take.

There are many choices of milk at the grocery store—whole, low-fat, skim, buttermilk. Which is best? Except for the amount of fat and calories, there is little nutritional difference between the different milks. Whole milk, from which none of the fat has been removed, contains about 3½ percent fat. Low-fat milk usually contains either 1 or 2 percent fat, and skim milk has had most of its fat removed. Removal of fat also removes vitamin A, so manufacturers usually add this vitamin to low-fat and skim milk. **Cultured buttermilk** is made by adding fermenting microbes to low-fat or skim milk. While lower in fat, salt is usually added for taste. So it is higher in sodium than **sweet milk** (unfermented milk). (See Chapter 23, "Chunky," for more information about buttermilk.) All milks are usually fortified with vitamin D.

For safety, milk is **pasteurized**, which means it is heated to kill disease-causing bacteria in it. After being pasteurized, milk is **homogenized**, meaning it is made homogeneous. The milk is forced through tiny openings that break the fat into very small particles that can stay uniformly distributed throughout the milk. Natural milk is said to be **unhomogenized**. In unhomogenized milk, the fat particles are lighter than the milk and float to the surface. (See Chapter 23, "Chunky," for more information about milk fat.)

Exercises

1. Which figure, A or B, shows milk with the greater amount of fat?

A

B

2. Study the figures of the children. If each child is lactose intolerant, which child, A, B, or C, is most likely to have a bellyache after eating the food shown?

A   B   C

Activity: CLUMPED

Purpose To show that milk has calcium.

Materials masking tape
marking pen
three 10-ounce (300-ml) plastic transparent
   glasses
measuring cup
¾ cup (189 ml) milk (fat content doesn't matter)
measuring spoons
2 teaspoons (10 ml) vinegar
½ teaspoon (2.5 ml) Epsom salts
3 spoons

Procedure

1. Use the tape and pen to label the glasses A, B, and C.

2. Pour ¼ cup (63 ml) of milk into each glass.

3. Add 1 teaspoon (5 ml) of vinegar each to glasses A and B.

4. Add the Epsom salts to glass B and stir until dissolved.

5. Do not add anything to glass C. This is the control.

6. Use two spoons, one for glass A and one for glass C. Stir the liquids one at a time, then scoop some of the liquid into the spoons. As you slowly pour the liquids back into the glasses, compare them, looking specifically for the presence of any clumping.

7. Repeat step 6, using glasses B and C.

Results   The milk in glass A contains clumps. The milk in glasses B and C has no clumps.

Why?   Milk contains calcium ions. Vinegar is an acid, which in the presence of calcium ions will cause protein in milk to coagulate, such as happened in glass A. Epsom salts reacts with the calcium ions in the milk in glass B and changes it to an insoluble calcium substance. If enough Epsom salts is added, no calcium ions remain in the milk. Because no calcium ions were left in glass B, the milk did not **curdle** (coagulate) when the vinegar was added. Glass C was a control used to compare the results of the other glasses. (The calcium content of milk varies very little in milks with different fat content, such as whole milk, low-fat milk, and skim milk.)

Solutions to Exercises

1. *Think!*

- Dairy milk comes from cows.

- Milk without any fat removed is called whole milk.

- Whole milk is about 3½ percent fat.

- Skim milk has little or no fat in it.

  Figure B shows milk with the greater amount of fat.

2. *Think!*

- Eating milk and other dairy products causes a bellyache in people who are lactose intolerant.

- Which of the foods shown are dairy products? Ice cream and yogurt.

- Yogurt is a dairy product, but because the lactose in yogurt has been predigested by bacteria, this food can be eaten by a person with lactose intolerance without causing pain.

- Ice cream generally causes a bellyache in a person who is lactose intolerant.

- Which child is eating ice cream?

Child A is most likely to have a bellyache after eating the ice cream.

23
Chunky

How Dairy Products Are Made

What You Need to Know

The milk that you drink, be it whole, nonfat, or skim, is sweet milk (unfermented milk). Sweet milk contains the sugar lactose. As milk ages, or if it stays unrefrigerated for too long, fermentation occurs. The bacteria that multiply in the warm milk cause the lactose to ferment. One of the products of the fermentation of lactose is lactic acid. Since acids taste sour, fermented milk is also called **sour milk**.

But sour milk isn't all bad. The fact that milk goes "sour" is what gives us yogurt and cheese. When enough lactic acid is produced, it causes casein, milk protein, to coagulate, forming white lumps called curds. Milk curd consists mainly of proteins and fats plus some minerals, vitamins, and lactic acid. Since most minerals and vitamins are water-soluble, most remain in the whey, the liquid part of the milk left when curds separate. Any acid added to milk will cause milk to curdle (separate into curds and whey).

Yogurt is milk that has been soured by two special bacteria, *Lactobacillus bulgaricus* and *Streptococcus thermophilus*. These microbes curdle the milk and give it a particular flavor. The curds and whey do not separate, but instead form a thick custardlike product.

Cheese is the curd of milk that has been specially prepared. There are hundreds of varieties of cheese, some hard and some

soft, some with a strong flavor and some mild. But all of these different cheeses could be made from the same milk. The different cheese characteristics are due to the types of microbes used to prepare them as well as the process by which the curds and whey are separated. Almost all kinds of cheeses have added salt. It may be mixed with curd or applied to the surface of a pressed form. The salt not only gives flavor, but helps prevent the growth of undesirable microbes.

Most cheeses are ripened. In ripening, microbes are added and given time to multiply, and the chemicals they give off flavor the cheese. For hard cheeses, such as Swiss or cheddar, the microbes are mixed throughout the mass of curd to be ripened. These cheeses can be made in very large sizes. Soft cheeses, such as Limburger and Camembert, are ripened by placing microbes on the surface of the cheese. These cheeses must be made in small sizes so the chemicals produced by the microbes can move from the surface through the cheese.

Cottage cheese and cream cheese are soft unripened cheeses. Cottage cheese is made from the curds of skim milk. Cream cheese is made from the curds of whole milk to which cream has been added.

Processed cheese is a mixture of ground cheese (usually cheddar and other hard varieties), along with special ingredients to give the cheese flavor and make it last longer. American cheese is a processed cheese known for its mild taste and ability to melt smoothly.

Cream is a combination of milk fat and milk. You can see cream form by allowing unhomogenized milk to stand. The mixture of fat globules and milk that floats to the surface of the milk is cream. Cream floats because the milk fat in it has less **density** (weight per volume) than milk. The cream you purchase in the store has specific amounts of fat. Heavy cream (whipping cream) contains at least 36 percent milk fat. Light creams have 18 to 30 percent fat, and half-and-half (a milk-and-cream blend) contains about 10 to 18 percent fat. Sour cream is light cream or half-and-half with added microbes that produce lactic acid. The acid gives the cream a sour taste and makes it thick. It is homogenized for a uniformly smooth texture. Sour cream has the fat content of the cream from which it is made.

Butter is milk fat that has been separated from cream. In the activity in this chapter, you learn how to make butter from cream.

Exercises

1. Which figure, A, B, or C, shows a way of changing the milk in the glass into curds and whey?

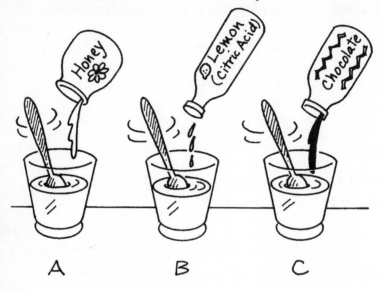

2. Two bottles of unhomogenized milk are allowed to stand. Which bottle, A or B, correctly shows where cream will form in the milk?

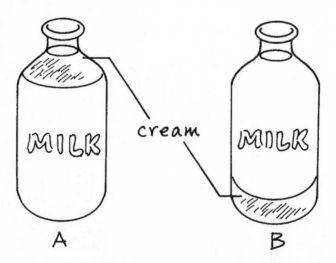

Activity: WHIPPED

Purpose To make butter from cream.

Materials ½ pint (250 ml) heavy (whipping) cream
timer
1-quart (1-liter) jar with a lid
cold tap water
1-pint (500-ml) plastic container with lid
spoon
⅛ teaspoon (0.6 ml) table salt (optional)

Procedure

*CAUTION: Only eat the products of this experiment if all uten-
sils are kept clean and instructions are carefully followed.*

1. Allow the cream to sit in its closed container for 30 minutes, but no longer, to reach room temperature.

2. Pour the cream into the jar and close the lid.

3. Shake the jar of cream vigorously until it is very thick.
   This will take 1 to 2 minutes.

4. Continue to shake the jar vigorously until clumps separate
   from the liquid.

5. Open the jar and pour out as much of the liquid as possible.

6. Wash the clumps remaining in the jar by filling the jar about one-fourth full with cold water.

7. Close the jar and shake vigorously about 10 times.

8. Open the jar and pour out as much of the liquid as possible. Discard this liquid.

9. Repeat steps 6 through 8 two or more times.

10. Pour the solid from the jar into the plastic container. Mix in the salt if you wish. Close the lid on the plastic container and refrigerate immediately.

11. After 1 hour or more, taste the solid in the plastic container.

Results   Shaking causes solids in cream to separate from the liquid. The solid you made tastes like butter.

Why?   Shaking the cream is called churning (shaking with a constant, vigorous motion). When the cream is churned, the fat globules break and the liquid fat helps cement other fat globules together. This mixture of fat forms a semisolid mass called butter. The liquid that it separates from is called buttermilk. In this experiment, the buttermilk is more of a **sweet buttermilk**, indicating that the milk is unfermented. (Cultured buttermilk, which is what you would buy at the store, is fermented skim milk. It is thicker and has a sour taste because of the presence of acids.)

The slight yellow color of your butter is due to carotene and other fat-soluble pigments. In commercial butter, artificial color is often added for appearance and salt is added for taste.

Solutions to Exercises

1. *Think!*

- Any acid added to milk will cause milk to curdle, producing solid curds and liquid whey.

- Which figure shows an acid being added to the milk?

  Figure B shows a way of changing milk into curds and whey.

2. *Think!*

- Cream is a mixture of milk fat and milk.

- The density of cream is less than milk, so it floats to the surface of unhomogenized milk.

- Which bottle shows cream on top of the milk?

  Bottle A correctly shows where cream forms in unhomogenized milk.

24
Good to Bad

Why Foods Spoil

What You Need to Know

The food you eat is also food for microbes, including bacteria and fungi. Some food fungi, such as yeast, are single-celled microbes. Other food fungi, such as mold, may start out as a microbe, but can grow to be multicelled and very visible. **Molds** have a fuzzy appearance and grow on the surface of damp or decaying foods. Bacteria and fungi do have beneficial effects on food. For example, bacteria are used to make buttermilk, molds are used in making cheeses such as Roquefort, and yeast is used in bread making. But microbes can also cause food to **spoil** (become unfit for use). Bacteria can make milk **blinky** (sour), yeast can make pickle juice develop a film, and black mold can grow on bread. Skimming away film or trimming away moldy areas and eating the remaining food is not advisable.

Spoilage is generally associated with any disagreeable change in food that can be detected with the senses of smell, touch, taste, or sight. Some spoilage occurs as a result of natural chemical reactions in the food, such as overripening, which causes the food to change color, lose flavor, and eventually become not good to eat. Cooking prevents this deterioration because high heat destroys most of the enzymes that cause the chemical reaction of ripening.

But most spoilage is due to the presence of microbes and the chemicals they produce. Microbes grow readily in foods that

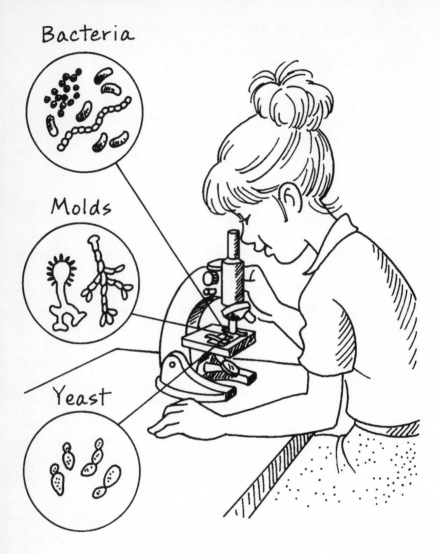

Bacteria

Molds

Yeast

contain protein, such as potatoes, rice, beans, meat, poultry, fish, and eggs. Most microbes require a warm, moist environment for survival.

It is important not to eat spoiled food because some microbes produce **toxins** (poisons). Eating foods with toxins would cause you to have **food poisoning**. The symptoms of food poisoning include sudden abdominal pain, nausea, vomiting, and diarrhea. The following are some helpful hints for preventing food poisoning:

• *Sanitation.* Keep your hands and nails clean. Wash your hands with soap and water after handling each food. Keep cooking utensils clean, washing them with soap and water after every use. Use different spoons for stirring different foods, especially raw foods such as meat and eggs.

• *Temperature.* Keep hot foods hot, 160°F (71°C) or higher, and cold foods cold, 40° to 45°F (4° to 7°C) or lower. Thaw frozen foods in the refrigerator instead of at room temperature, and do not refreeze thawed foods. While freezing stops bacterial growth, it does not kill all bacteria or destroy bacteria and toxins in food. Thawing and refreezing gives more opportunities for bacteria and toxins to build up in food.

The three most common food poisonings are botulism, staphylococcic food poisoning, and perfringen food poisoning. The bacterium *Clostridium botulinum* causes **botulism**, which is the most serious kind of food poisoning. If low-acid foods, such as meats and vegetables, are not properly heated when canned, *C. botulinum* can grow. Improper heating is more likely to occur in home-canned than commercially canned foods. Never taste home-canned vegetables and meats until they have been cooked at a boiling temperature for at least 20 minutes or as directed in a home-canning guide. Botulism can be fatal.

Staphylococcic food poisoning, commonly called "staph," is the most common food poisoning and is caused by ingesting foods containing toxins produced by *Staphylococcus aureus*. A wide variety of foods are good breeding areas for this disease-producing bacterium, such as custards, cream-filled desserts, tuna, chicken, potato salad, and gravy. Foods are generally contaminated by food handlers, and large amounts of toxins are produced in a few hours if the food is not refrigerated.

The bacterium *Clostridium perfringens* causes **perfringens food poisoning**, which is a mild form of food poisoning. It is called the cafeteria bug because it is common where bulk quantities of foods are served, and the cold foods and hot foods have both been kept warm (at room temperature) for an extended period.

Other microbes that contaminate food and cause illness are *Escherichia coli,* commonly called *E. coli,* and hemorrhagic *E. coli. E. coli* is present in unpasteurized milk and undercooked meats, especially hamburger and all-beef hot dogs. Illness caused by *E. coli* bacteria is especially dangerous in the very young, the elderly, and the feeble. To prevent illness, follow the general rules of safety and always drink pasteurized milk and cook meat thoroughly to kill the bacteria. With hamburgers, this means until the center is brown.

Salmonellosis is an infection caused by bacteria of *Salmonella.* The food poisoning is commonly called "salmonella." Unlike other kinds of food poisoning, salmonellosis is caused by the bacteria themselves, not toxins. Some of the ways to prevent this disease are to cook eggs, poultry, meat, and shrimp thoroughly. Also practice strict sanitation when handling any of these raw foods.

Exercises

1. Which figure, A or B, shows a process that could result in food poisoning?

2. Which type of hamburger, rare or well done, is more likely to contain *E. coli* bacteria?

Activity: JUST RIGHT

Purpose To determine a good temperature for the growth of bread mold.

Materials masking tape
marker
two 1-quart (1-liter) resealable bags
2 slices of white bread
tap water
2 cotton balls

NOTE: A refrigerator is necessary for this activity.

Procedure

1. Use the tape and marker to label the bags Warm/Wet and Cold/Wet.

2. Place one slice of bread in each of the bags.

3. Moisten the cotton balls and place them in the bags.

4. Place the bag labeled Warm/Wet in a dark, warm spot, such as a cabinet at room temperature.

5. Place the bag labeled Cold/Wet in the freezer.

6. Observe the bread slices through the plastic bags daily for 14 or more days.

7. At the end of the experiment, discard the bags and their contents.

Results   A black, hairy-looking growth is seen on the bread in the Warm/Wet bag. No growth is seen on the bread stored in the freezer.

Why?   Microbes grow best in warm, moist environments. In this experiment, the moist cotton supplied the moisture needed for microbial growth, and at room temperature, 65° to 75°F (18° to 24°C), bread mold grew on the bread.

The temperature of a freezer is approximately 0°F (–18°C). Microbes also need water, and since water is ice at this temperature, most microbes cannot grow in the freezer. Not only does freezing inhibit microbial growth, but over 50 percent of some kinds of microbes in food may be killed during freezing.

Solutions to Exercises

1. *Think!*

- Figure A shows a frozen food thawing on a sunny table.

- Thawing food at room temperature encourages the growth of microbes in the food.

- Figure B shows a boy washing his hands.

- Keeping your hands clean is a good way to keep the food you handle clean.

   Figure A shows a process that could result in food poisoning.

2. *Think!*

- *E. coli* is present in undercooked meat.
- Cooking meat thoroughly is the best way to kill *E. coli.*

A rare hamburger is more likely to contain E. coli *bacteria than a hamburger that is well done.*

25

Long-Lasting

How to Make Food Last

What You Need to Know

Christopher Columbus (1451–1506), an Italian explorer, discovered the New World, but his intent was to find a new sea route to the spice lands of Asia. Spices were valuable then as they had been for centuries, not only for adding flavor to foods but in **preserving** (making something last) them. In order to make food last longer, the food must be changed to inhibit the growth of microbes. Some spices, such as cloves, inhibit the growth of bacteria.

Spicing food is only one of many past and present ways of food preservation; others include cooking, drying, canning, chilling and freezing, freeze-drying, and food additives. Cooking, the oldest of the food preservation methods, not only kills bacteria but makes some nutrients more digestible. Starch, for example, must be cooked in order for your body to digest it. While cooking makes proteins, fats and carbohydrates in foods more digestible, it can reduce or destroy minerals and vitamins. But as with all methods of food preservation, the good outweighs the bad.

Even before the Native Americans hung strips of buffalo meat in the sun and wind to dry, people dried meats and other foods. They didn't know they were producing an unacceptably dry environment for microbes, but they did know it kept their food from spoiling. While drying **dehydrates** (removes water from) food and makes it last longer, it does affect its taste—usually

unfavorably. But some dehydrated foods, such as raisins (dehydrated grapes) and jerky (dehydrated beef), are sold as snack foods. Drying almost always causes some loss of nutritional qualities.

Salting is another way of preserving foods by dehydration. This method is widely used today, especially for meat, fish, pickles, and sauerkraut. But the downside is that these foods are very high in sodium. Sodium nitrate, a form of salt used to preserve bacon and ham, when overheated will combine with the amino acids in protein and produce **carcinogens** (cancer-causing substances) called nitrosamines.

In the 1790s the French government offered a prize of 12,000 francs for a food preservation method that could be used to feed a large, constantly moving army. Nicolas Appert (1749–1814), a French cook, was inspired by this offer and began experimenting. After 14 years, he devised a method for heating foods sealed in glass bottles and won the prize in 1809. The method, still used today, is called canning. Appert had no understanding of microbes, but observed that if he sealed the jars tightly, and heated them in boiling water, the food did not spoil. The food remained edible until the jars were opened. Canning techniques improved quickly. By the time of the American Civil War (1861–1865), canning was the major method of food preservation. Modern improvements have made canning an inexpensive and popular method of food preservation.

Chilling and freezing foods may be the very oldest methods of preserving food. Before modern refrigeration, snow and ice were wrapped in insulating materials such as straw to keep them from melting so they could be used to preserve foods. Another method of chilling foods was to store them in cool wells and cellars. Chilling and freezing use low temperatures to prevent microbial growth; and freezing like heating kills some microbes. The colder the food is kept, the slower the growth of microbes. Freezing protects nearly all of a food's nutrients and flavor. As soon as the frozen food is warmed, the surviving microbes become active again and the food must be used quickly to avoid spoilage.

In freeze-drying, foods are frozen in a **vacuum** (a space from which almost all the air or gas has been removed), then their ice crystals are **vaporized** (changed to a gas). Freeze-dried food contains only 1 to 8 percent of its original moisture. Almost anything can be freeze-dried. Freeze-dried foods have a long **shelf life** (the time from a food's preparation for storage until it spoils). When **rehyhdrated** (water is restored to dehydrated materials), the food is said to be **reconstituted**. Reconstituted freeze-dried food has good flavor and texture, and a minimum of nutrients are lost. But because of the cost, only a few foods are freeze-dried, such as coffee and camp rations. Some of the food eaten by astronauts in space is freeze-dried.

Food additives are used to maintain freshness and extend the shelf life of packaged foods. Many people think additives are mysterious chemicals that may have unwanted side effects. However, many **preservatives** (food additives that increase the shelf life of foods) are natural food substances, such as sugar, salt, sodium ascorbate (vitamin C), and vitamin E. Sodium propionate, a chemical naturally present in Swiss cheese, is commonly added to baked goods such as bread to slow the growth of mold.

BHA (butylated hydroxyanisole) and BHT (butylated hydroxytoluene) are chemical antioxidants that help keep fats and oils from combining with oxygen and becoming **rancid** (having a strong offensive smell or taste). They are used to prolong the shelf life not only of foods containing fat or oil but also of other foods, such as cereals, rice, and frozen fruits.

Exercises

1. Which figure, A, B, or C, shows the effect of freezing on microbes in food?

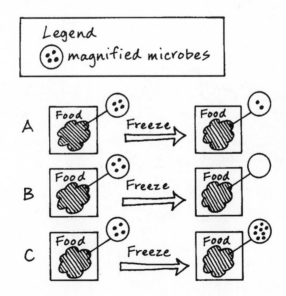

2. Study the figures A and B and identify the food preservation method each represents.

Activity: CHILLING

Purpose To determine the effect of freezing on the crisp-ness of foods.

Materials tape
          pen
          four 1-pint (500-ml) resealable freezer bags
          large lettuce leaf
          carrot
          tap water
          paper towel

NOTE: A freezer and a refrigerator are needed for this activity.

Procedure

1. Use the tape and pen to label two of the bags R and the other two bags F.

2. Wash the foods with running water and pat them dry with the towel.

3. Break the lettuce leaf in half, placing one piece in an R bag and the other in an F bag.

4. Press the bags to remove as much air as possible without crushing the food.

5. Seal the bags. Place the F bag in the freezer and the R bag in the refrigerator.

6. Repeat steps 3 through 5, using the carrot.

7. After one day, remove the bags from the freezer and place them in the refrigerator with the other bags to thaw.

8. After another day, remove all the bags from the refrigerator.

9. Test the crispness of each of the foods by feeling them with your hand. Try to break the food.

Results The foods in the freezer have lost their crispness and become limp.

Why? As wonderful as freezing may seem, it is not suitable for all foods. Those containing large amounts of water and crisp foods eaten raw, such as lettuce and carrots, have an unpleasant texture when thawed (see Chapter 13, "Icy"). While freezing causes raw carrots to lose their crispness, it is still a great way to preserve this food because frozen carrots taste good when cooked.

Solutions to Exercises

1. *Think!*

- Freezing temperatures help to prevent microbial growth and may kill 50 percent or more of some kinds of microbes, but generally not all microbes are killed.

- Which figure shows a decrease of some but not all microbes?

  Figure A shows the effect of freezing on microbes in food.

2a. *Think!*

- Figure A shows a sealed jar of food being heated.

- Which food preservation method involves heating food in a sealed container?

  Figure A shows the canning process.

b. *Think!*

- Figure B shows grapes lying in the sun.

- What effect does the sun have on food? It dries food.

  Figure B shows dehydration, the drying method.

Appendix

Anthocyanin Indicator

Materials   12 leaves of red cabbage
blender
measuring cup
distilled water
large strainer
large bowl
1-quart (1-liter) jar with lid
wide masking tape
marking pen
adult helper

Procedure

1. Tear the cabbage leaves into small pieces and put them in the blender.

2. Use the cup to add enough water to cover the cabbage in the blender.

3. Ask an adult to blend the water and cabbage.

4. Hold the strainer over the bowl and empty the contents of the blender into the strainer.

5. Pour the cabbage juice from the bowl into the jar, and discard the solid pieces of cabbage left in the strainer.

6. Use the tape and marking pen to label the jar Anthocyanin Indicator.

7. To prevent spoilage, store the Anthocyanin Indicator in a refrigerator until needed for the activity in Chapter 17, "Pucker Up!"

Glossary

abdomen The part of the body containing the coiled small intestine; the belly.

absorbed Taken in or soaked up, as a sponge absorbs liquids.

acid A chemical that has a pH of less than 7, tastes sour, and neutralizes a base; a chemical formed by carbohydrates in your mouth that attacks tooth enamel; a chemical in gastric juices that kills most bacteria in the stomach and small intestine.

activate To make more active.

aleurone The layer of cells below the bran layers of wheat grain that contain the highest-quality protein.

alimentary canal The tubular passage in the body of humans and animals through which food passes and is digested.

all-purpose flour Flour made from a mixture of hard and soft wheat.

amino acids The building blocks of proteins, made of carbon, hydrogen, oxygen, and nitrogen. There are 20 types in the human body, from which all protein is built.

anthocyanins Red, purple, and blue plant pigments used as natural food colors and as an indicator.

anthozanthins White plant pigments used as natural food colors.

antibodies Disease-fighting proteins in the body.

antioxidant A substance that inhibits oxidation.

anus   The end of the rectum that exits the body, through which solid wastes leave the body.

artificial   Man-made.

aspartame   A calorie-free artificial sweetener made from the combination of two amino acids, aspartic acid and phenyl-alanine.

atom   The smallest unit of a molecule.

attract   To pull toward each other.

bacteria   A one-celled microbe found all around us.

baking powder   A leavening agent that is a mixture of baking soda, a dry acid, and cornstarch.

baking soda   A leavening agent that has the chemical name sodium bicarbonate and the common name soda.

balance   Equal amounts; one of the three key words for good nutrition.

basal metabolism   The amount of energy used by the body while resting and/or fasting to carry out basic functions.

base   A chemical that has a pH of greater than 7, tastes bitter, and neutralizes an acid.

batter   A flour-and-liquid mixture that can be poured or dropped from a spoon.

beat   To mix batter by stirring as with a spoon or electric mixer.

beta-carotene   A natural yellow food dye that can be converted by the body into vitamin A; thus, "provitamin A" accompanies it on food labels.

bile   A liquid made by your liver and stored in your gall bladder which mixes with fats in your small intestine, breaking the fats into tiny globules so enzymes can digest them; an emulsifier.

blinky   Sour, as spoiled milk.

bloat   To swell up.

bloodstream   The blood flowing throughout your body.

blood vessels   Tubes that transport blood throughout your body.

body mass index (BMI)   Body weight in kilograms divided by the square of height in meters. Average BMI for boys is 22 to 24, and for girls it is 21 to 23.

bolus   The food ball prepared in the mouth and swallowed.

bond   A connection between atoms.

botulism   The most serious type of food poisoning, caused by the toxins produced by the bacterium *Clostridium botulinum.*

bran   The nutritious layers forming a protective covering beneath the chaff of a wheat grain.

bromelin   An enzyme in pineapples that digests protein such as that found in gelatin.

bud pore   The opening at the top of a taste bud through which liquids can enter.

buds   Parts of a stem that develop into other stems or flowers.

butter   A dairy product; milk fat that has been separated from cream by churning.

buttermilk   An acidic liquid; the liquid left when butter is separated from churned cream.

calorie   The unit of measure for food energy. The abbreviation for nutrient calories is Cal.

carbohydrates   Macronutrients. Chemicals made of carbon, hydrogen, and oxygen and organized into two groups, simple and complex, that are the body's most important source of energy; nutrients mainly obtained from plants.

carcinogens Cancer-causing substances.

carotenes Vivid yellow and orange plant pigments used as food colors.

casein A protein in milk that coagulates when mixed with rennin or acid.

cells Small building blocks of living things.

cellulose A complex carbohydrate in your diet; a chemically indigestible fiber; needed for proper removal of feces from the body; commonly called crude fiber; good food sources are cereals, breads, potatoes, grains, peas, and beans; provides structure for plants.

chaff The strawlike covering around a wheat grain.

cheese A dairy product; the curd of milk that has been specially prepared.

chemical digestion The breaking apart of long chains of food molecules into smaller units of combined or separate molecules.

cholesterol A type of fat that can clog blood vessels.

churn To move about; to shake with a constant, vigorous motion as in the churning of cream to make butter.

chyme The liquid food mixture produced in the stomach by the mixing of churned food with gastric juice.

clot To form a mass or lump; mass or lump.

coagulate To thicken and form clumps.

coenzyme A chemical helper needed for some enzymes to do their job, such as vitamins.

collagen An insoluble protein in meat that causes the meat to be tough. Makes up one-third or more of your body's protein. A molecule of collagen is made of three protein chains loosely wound together and held by hydrogen bonds.

collagen fiber   A type of connective tissue in the body of animals that is made of a bundle of many macrofibrils and is very strong and very resistant to stretching.

colloid   A homogeneous mixture of particles suspended in a gas, liquid, or solid.

colon   The large intestine, except for the rectum.

complementary proteins   A combination of incomplete proteins that provide the nine essential amino acids.

complete protein   A protein that contains all nine essential amino acids in the right amounts needed by the body; sources are poultry, fish, eggs, meat, and dairy products.

complex carbohydrates   Chains of many saccharide molecules; commonly called polysaccharides; starch and dietary fiber.

concentration   A measure of how closely packed materials are.

condensation   The process of changing a gas to a liquid; requires a loss of heat energy.

condense   To change from a gas to a liquid.

connective tissue   Tissue in the body made of collagen fibers which binds together and supports the tissues of your body.

contraction   A squeezing together, as a muscle contraction resulting in a thickened and shortened muscle.

cream   A dairy product; a combination of milk fat and milk.

crude fiber   See **cellulose**.

crystal   A solid with its atoms arranged in a definite geometric shape.

cultured buttermilk Buttermilk made by adding fermenting microbes to low-fat or skim milk.

curd Coagulated milk protein.

curdle To coagulate; to separate into curds and whey.

dairy products Cow's milk and milk products such as butter, cheese, cream, and yogurt.

deactivate To make less active.

dehydrate To remove water; to dry out food or body cells.

dehydration An excess loss of water from your body.

denaturing The process of changing protein from its natural form.

dense Having materials close together.

density Weight per volume.

dental caries The eating away of tooth enamel by an acid produced as a result of a chemical reaction between sugars and other carbohydrates and normal bacteria in the mouth.

dietary fiber A part of food from plants; chemically indigestible complex carbohydrate that is needed for body functions such as proper removal of feces from the body and to lower cholesterol. See also **cellulose** and **pectin**.

diffuse To spread out evenly in all directions.

digestion The changing or breaking down of food by organisms, both mechanically and chemically, into particles.

digestive juices Liquids in the digestive system that digest food.

digestive system The parts of your body that work together to change the food you eat into particles small enough to be taken in by your cells; made up of the alimentary canal along with the salivary glands, pancreas, liver, and gall-bladder.

disaccharide A simple carbohydrate made of two sugar molecules and commonly called a double sugar; includes lactose, maltose, and sucrose.

dissolve To break apart and spread out in a substance.

dough A flour-and-liquid mixture that can be shaped by rolling or kneading.

durum wheat An especially hard wheat, the flour of which is used to make pasta.

dyes Coloring materials.

ectomorph Body type characterized by an elongated and slender build.

electrolytes Ions formed in body fluids.

electron A negatively charged particle spinning around the positive center of an atom.

emulsifier A substance that prevents an emulsion from separating.

emulsion A mixture of two liquids that do not dissolve in each other, such as oil and water. One of the liquids (oil) is suspended (hanging) in little drops in the other liquid (water). If allowed to stand, the liquids in a temporary emulsion separate.

endomorph Body type characterized by a rounded body with short arms and legs.

endosperm The starchy interior of a wheat grain that is used to make white flour.

enriched Having nutrients added to replace those lost during processing of food.

enzymes Proteins in living things that control chemical reactions by causing the reaction or changing their speed; their job can be to put together or take apart molecules.

epiglottis The flap of cartilage that closes the opening to the trachea to make sure the bolus enters the esophagus.

esophagus The muscular tube connecting the pharynx and the stomach.

essential A term used with some nutrients to indicate that the body cannot make them, so they must be part of the diet.

essential amino acids The nine amino acids that your body cannot make in large enough amounts to satisfy the nutritional requirements for good health and must be supplied by the food you eat.

essential fatty acids (EFAs) Fatty acids that the body cannot make and must be part of your diet; found in foods such as sunflower seeds, walnuts, leafy green vegetables, and corn, canola, safflower, soybean, and sunflower oils.

ethanol Drinking alcohol, produced by the fermentation of sugar by yeast.

ethylene gas A plant hormone that encourages fruit ripening.

evaporate To change from a liquid to a gas.

evaporation The process of changing from a liquid to a vapor at the surface of a liquid. Heat energy is required for this change.

fasting Not eating.

fats Macronutrients. Greasy substances stored in animals' and plant cells; a nutrient that is used as an alternative energy source when glycogen is used up; triglycerides.

fat-soluble Dissolves in fat.

fatty acids   Chemicals found in animal and plant fat, composed of carbon, hydrogen, and oxygen. Used in the formation of fat.

feces   Solid body waste.

fermentation   A chemical reaction in which microbes growing in the absence of air cause changes in food.

flavor   Taste, including those associated with smell, such as strawberry.

florets   Many tiny flowers that make up the head of a flower vegetable, such as broccoli and cauliflower.

food   Any animal or plant substance taken in by living things that is used to provide energy, and promote growth and other life-supporting processes.

Food and Drug Administration (FDA)   An agency of the U.S. Department of Health and Human Services that tries to ensure that our foods are safe.

Food Guide Pyramid   A diagram that shows the five basic food groups and how much of each food group you should eat each day.

food poisoning   Any of several diseases caused by toxins in food, such as botulism and staphylococcic food poisoning.

fortified food   A food product containing one or more added nutrients not naturally present in the product.

fructose   A simple sugar found in fruits and honey.

fruit   The part of the plant that contains the seeds.

fungus   A microbe or multicelled organism such as mold and yeast.

galactose   A simple sugar found in combination with other simple sugars in dairy products.

gallbladder   The body organ where bile is stored and released into the small intestine when fatty foods are present.

gastric juices   Digestive juices produced by the stomach that partially digest proteins.

gel   A semisolid colloid, like gelatin desserts, that is more solid than a sol.

gelatin   A gummy water-soluble protein obtained from animal tissues which is used in cooking to turn a liquid into a gel.

gelling   The process of changing a sol into a gel.

germ   The vitamin- and mineral-rich embryo of a wheat grain.

globular protein   Compact, rounded, coiled chains of amino acids that generally are water-soluble; found in egg whites.

glucose   A simple sugar in blood, called blood sugar; a fuel molecule for respiration.

gluten   A tough, stretchy protein in batter and dough.

glycerol   A chemical found in animal and plant fat, composed of carbon, hydrogen, and oxygen. One molecule of glycerol plus three molecules of fatty acids forms a fat molecule.

glycogen   A polysaccharide that is the form of carbohydrates stored in animals; called animal starch.

grains   Edible starchy fruits or seeds from various grasses; also called cereals.

GRAS   Generally Recognized as Safe, a list of safe food additives prepared by the FDA.

hard water   Water rich in the minerals calcium, magnesium, and/or iron.

hard wheat   Wheat containing a high amount of proteins, the flour of which is used to make bread.

hemoglobin   The red substance in blood that carries oxygen to cells.

hives   Intensely itchy areas on the skin, caused by such things as being sensitive to some artificial dyes in foods.

homogeneous mixture   A mixture in which particles of two or more substances spread evenly so that the mixture has the same composition throughout.

homogenized   Made homogeneous.

hormones   Messenger chemicals that are made in one part of a plant or animal and move to another part via plant and body fluids, where they cause a specific response in cells and tissues.

hunger   The body's physical need for food.

hydrogen bond   A loose bond, such as that formed by the attraction between the positive hydrogen end of one water molecule and the negative oxygen end of another water molecule.

hypertension   High blood pressure.

hypothalamus   A part of the brain that sends signals that make you feel thirsty or hungry.

immune system   The group of body parts that work to make the body resistant to diseases.

incomplete protein   A protein that does not contain all the essential amino acids; found in vegetables, especially legumes.

indicator   A pigment that changes color depending upon whether it is in the presence of an acid or a base.

inflexible   Not bendable.

ingest   To take, as food, into the body.

inhibit   To decrease or stop an action.

inorganic substances   Substances that do not contain carbon or come from living things.

input   To be taken in.

insoluble   Won't dissolve.

insulate   To cover with a material that reduces the passage of heat.

ion   A charged particle made of a single atom or group of atoms with a positive or negative charge.

kidneys   Organs that filter liquid waste from your blood to form urine and help to regulate the amount of sodium and water content in the body.

knead   To mix dough by pressing, folding, and stretching it.

lactase   An enzyme in your small intestine that digests lactose, breaking it down into glucose and galactose.

lactose   A disaccharide made of glucose and galactose, forming what is commonly known as milk sugar.

lactose intolerance   A disorder in people who have no lactase or too little lactase to digest the milk sugar lactose.

large intestine   The 5-to-6-foot (1.5-to-2-m) -long U-shaped tube of the alimentary canal between the small intestine and the rectum, where feces is formed.

leaven   To inflate with leavening gases.

leavening   The process by which gases make dough and batters rise.

leavening agent   A substance that produces leavening gases, such as baking soda, baking powder, and yeast.

leavening gases   Gases that leaven baked products and make them light and fluffy, such as carbon dioxide, air, and steam.

legume   A plant that bears seeds in pods, such as peas, beans, and peanuts; excellent sources of incomplete protein.

lipase   A fat-digesting enzyme.

liver   The body's largest internal organ, weighing 3 to 4 pounds (1.4 to 1.8 kg) in an adult. Part of its job is to store vitamins B_{12} and the fat-soluble vitamins A, D, E, and K. Bile, a green liquid made by the liver, helps to digest fat in the small intestine.

macrofibril   A bundle of many microfibrils.

macrominerals   Minerals needed by the body in large amounts; calcium, phosphorus, and magnesium.

macronutrients   Nutrients needed by the body in large quantities; water, carbohydrates, fats, and proteins.

maltose   A disaccharide made of two molecules of glucose, found in sprouting grain.

mammals   Animals that produce milk to nurse their young.

mechanical digestion   The physical breaking apart of food into smaller pieces.

mesomorph   Body type characterized by a muscular, medium build.

metabolism   Any one or the total of all chemical and physical processes by which the body uses food to release energy and uses the energy to build and repair body tissues.

microbes   Tiny living things visible only under a microscope, such as bacteria, molds, and yeast; also called microorganisms.

microfibril   A bundle of many collagen molecules.

microminerals Minerals needed by the body in small amounts; also called trace minerals; sodium, potassium, chloride, iron, zinc, iodine, copper, manganese, fluoride, chromium, selenium, molybdenum, arsenic, boron, nickel, and silicon; trace minerals.

micronutrients Nutrients needed by the body in small quantities; vitamins and minerals.

milling The process by which grains are ground and sifted into separate parts to make flour.

minerals Micronutrients. Nutrients that help the body use other nutrients; inorganic substances from water and soil that are essential to the functioning of your body.

moderation Not extreme—not too much and not too little; one of the three key words for good nutrition.

mold A fuzzy-looking fungus that grows on places such as damp or rotting food.

molecules The smallest particles of a substance that keep the properties of the substance and are made up of atoms.

monosaccharide A simple carbohydrate made of one sugar molecule and commonly called a simple sugar; includes fructose, galactose, and glucose.

monounsaturated triglyceride A triglyceride with one double bond between the carbons of at least one of its fatty acids.

myoglobin The substance that stores oxygen in muscles.

nasal passage The opening from your nose into your throat.

natural Found in nature.

nerves Bundles of cells that the body uses to send messages to and from the brain and spinal cord.

neutral No charge. Having a pH of 7 and thus being neither acidic nor basic, and tasting neither sour nor bitter.

neutralization   The reaction of an acid and a base to produce two neutral substances, salt and water.

nut   Fruits with a hard shell that contain only one seed, such as walnuts, pecans, almonds, and Brazil nuts.

nutrients   The materials in food that your body needs to grow, have energy, and stay healthy.

nutrition   The science that deals with the processes by which living things take in and use food.

nutritionist   A scientist trained in the science of nutrition.

obesity   Overweight.

oil   An unsaturated triglyceride that is liquid at room temperature; found in plants.

organic substances   Substances that contain carbon and come from living things.

organism   Any living thing, such as a plant or an animal.

organs   Groups of tissue that perform the same job, such as the liver and gall bladder.

osteoporosis   A medical condition in which calcium is lost from bones, causing the bones to become easily broken.

output   To be lost or taken out.

oxidation   The combination of a substance with oxygen.

oxidized   Combined with oxygen.

pancreas   The organ located near the small intestine which releases pancreatic juice into the small intestine.

pancreatic juice   Digestive juice produced by the pancreas and released into the small intestine which digests many of the molecules of food in the small intestine that are too large to be broken apart by bile.

papain   An enzyme extracted from papaya, a tropical fruit, and used as a meat tenderizer.

papillae   Red bumps on the tongue that contain taste buds.

pasteurized   Heated to kill disease-causing bacteria.

pectin   A plant complex carbohydrate; a water-soluble dietary fiber; found in a layer between the walls of cells that touch each other; helps to bind the cells together.

Percent Daily Value (DV)   The percentage of the suggested daily amount of a nutrient in a food serving based on a 2,000 calorie diet. Over the course of a day, the Percent DVs of nutrients in foods consumed should add up to about 100 percent.

perfringens food poisoning   A mild form of food poisoning caused by the toxins from the bacterium *Clostridium perfringens;* commonly called the cafeteria bug because it is common where bulk quantities of food are served.

peristalsis   Waves of muscle contractions inside the esophagus and intestines to move food through the digestive system.

perspiration   Salty water released through pores in the skin; commonly called sweat.

pH scale   A scale used to measure and compare the amount of acid or base in a food.

pharynx   The throat.

phloem tubes   Plant tubes that transport sap containing dissolved nutrients, mainly sucrose, throughout the plant; the outer circle seen in the cross section of a carrot.

photosynthesis   A process by which green plants use light energy to change carbon dioxide and water into glucose and oxygen.

pigment   A natural substance that gives color.

polar molecules Molecules that have a positive charge on one side and a negative charge on the other side.

polysaccharides See **complex carbohydrates**.

polyunsaturated triglyceride A triglyceride with two or more double bonds between the carbons of at least one of its fatty acids.

pores Tiny holes, as in your skin.

preservatives Food additives that increase the shelf life of foods.

preserve To make something last, as in preserving food to prevent spoilage.

protein A macronutrient. Chains of amino acids which are needed for growth and repair of your cells.

proton A positively charged particle in an atom's center.

rancid Having a strong offensive smell or taste.

Recommended Daily Allowance (RDA) The daily amount of the different food nutrients deemed adequate for healthy individuals by the Food and Nutrition Board of the National Research Council, a branch of the National Academy of Sciences.

reconstitution The process of rehydrating dried food.

rectum The final section of the large intestine, about 7 inches (17.5 cm) long, where feces is stored until it exits through the anus.

refined Separated and freed from impurities.

rehydrate To restore water to dehydrated materials.

rennin An enzyme taken from the lining of a calf's stomach which is used in cheese making.

respiration A chemical change in cells in which energy is released by the oxidation of glucose; includes breathing, which brings in the oxygen for oxidation.

root vegetables Edible taproots of plants, including carrots, beets, radishes, yams, sweet potatoes, and turnips.

saccharide Sugar.

saccharin A calorie-free artificial sweetener.

saliva A liquid in the mouth released by the salivary glands that kills bacteria, and softens, lubricates, and partially chemically digests starch in the food you eat.

salivary glands Glands located in and near the mouth, which produce saliva.

salmonellosis An infection caused by bacteria of *Salmonella;* commonly called "salmonella."

salt Sodium chloride, a mineral composed of sodium and chloride; common name for table salt.

sap A liquid in plants containing water and dissolved nutrients.

saturated fat See **saturated triglyceride**.

saturated triglyceride A triglyceride with single carbon-to-carbon bonds; commonly called saturated fat; found in animals and plants; examples are bacon grease, butter, lard, chocolate, and coconut.

selectively permeable membrane A barrier that allows some, but not all, materials to pass through.

shelf life The time from a food's preparation for storage until it spoils.

shortening The term used for the fat used in making dough, so named because it shortens gluten strands.

simple carbohydrates   Monosaccharides and disaccharides. Sugars that are more easily broken apart by digestion than are complex carbohydrates; tend to taste sweet, form crystals, and dissolve in water.

simple sugars   See **monosaccharides**.

small intestine   The part of the alimentary canal between the stomach and the large intestine, about 13 to 17 feet (4 to 6 m) long and about 1 to 1½ inches (2.5 to 4 cm) in diameter, where most digestion in the body occurs.

soap scum   A waxy material that does not dissolve in water. Produced by mixing soap and hard water.

soft palate   The soft back part of the roof of your mouth that moves up to close off the nasal passage when you swallow.

soft water   Water that has little if any of the minerals calcium, magnesium, or iron.

soft wheat   Wheat containing a low amount of proteins, the flour of which is used to make cakes.

sol   A colloid composed of solid particles suspended in a liquid.

soluble   Able to dissolve in a substance.

sour milk   Fermented milk.

spinal cord   Bundle of nerves running from the brain down the back.

spoil   To become unfit for use.

staphylococcic food poisoning   A common food poisoning caused by ingesting foods containing toxins produced by *Staphylococcus aureus;* commonly called "staph" food poisoning.

starch   A part of food from plants; a digestible complex carbohydrate in your diet; good food sources are cereals, breads, potatoes, grains, peas, and beans; main storage form of carbohydrates in plants.

steam   Water in the form of water vapor.

stem   A plant structure that bears leaves and has buds that become a stem or a flower.

stomach   A pouchlike part of the alimentary canal between the esophagus and the small intestine; the part of the digestive system where the bolus is churned into small pieces and mixed with gastric juices, which change the solid food to liquid.

succulent   Juicy.

sucrose   A disaccharide made of two simple sugar molecules, glucose and fructose, commonly known as table sugar or granulated sugar.

sugar   A simple carbohydrate; a saccharide.

supermedicinal   Having exceptional healing properties.

supplement   A source of nutrients taken in addition to foods.

sweat   A common name for perspiration.

sweet milk   Unfermented milk.

synthetic   Not natural; made by the combination of chemicals.

taproot   A single large plant root that not only helps anchor the plant in the ground but serves as an underground storage area for food and water for the plant.

taste buds   Groups of cells located on the tongue and on the roof and back of the mouth that are responsible for the sense of taste.

tissue   A group of similar cells working together to perform a function, such as muscle tissue.

tooth enamel   The hard protective surface on teeth.

toxins Poisons produced by some microbes.

trace minerals See **microminerals**.

trachea The tube through which air passes into lungs; also called the windpipe.

translucent Capable of letting some light pass through.

triglycerides Fats; a combination of three molecules of fatty acids plus one molecule of glycerol.

tuber The swollen underground stem of a plant that stores reserve food and water for the plant, such as Idaho potatoes and taro.

unhomogenized Not homogenized, as natural milk.

unit cells The basic building blocks of crystals.

U.S. Department of Agriculture (USDA) A federal agency whose aims include stamping out hunger and malnutrition.

unsaturated fat See **unsaturated triglyceride**.

unsaturated triglyceride A triglyceride with one or more double bonds between the carbons of the fatty acids; commonly called an unsaturated fat; a good source is plants; examples are corn, peanut, and olive oils.

unstable Changeable.

urine Liquid body waste.

vacuum A space from which almost all the air or gas has been removed.

vapor Gas.

vaporize To change to a gas.

variety Different kinds; one of the three key words for good nutrition.

vegetables   Any part of a plant that is eaten: roots, stems, leaves, fruits, nuts, and seeds.

vegetarian   Someone who does not eat meat; some do eat animal products such as milk, cheese, eggs, and honey.

viscous   Thick.

vitamins   A micronutrient. A nutrient that helps the body use the other nutrients; organic chemicals the body needs for normal growth and metabolism.

water   A macronutrient. The most abundant nutrient in your body; necessary for all body functions.

water-soluble   Dissolves in water.

whey   The liquid part of milk without the curd.

wilt   To become limp.

xylem tubes   Plant tubes carrying sap with dissolved nutrients from the soil to all the plant parts; the inner circle seen in the cross section of a carrot.

yeast   A leavening agent that slowly digests sugars and starches, producing carbon dioxide and ethanol; a microbe; rich in B vitamins and iron.

yogurt   A dairy product; milk that has been soured and caused to curdle by two special bacteria, *Lactobacillus bugaricus* and *Streptococcus thermophilus*.

Index

Abdomen:
definition of, 82, 199
Absorbed:
definition of, 36, 199
Acid:
cultured buttermilk, 176
definition of, 81, 126, 199
in foods, 125, 130, 176
lactic, 171
mouth, 107
neutralization of, 125
taste of, 125
vinegar, 169
Activates:
definition of, 132, 199
Aleurone:
definition of, 149, 199
Alimentary canal:
definition of, 79, 199
All-purpose flour:
definition of, 150, 199
American cheese, 173
Amino acids:
definition of, 27, 199
essential, 28, 206
Anthocyanins:
definition of, 120, 199
indicator, 127–129, 197–198
Anthozanthins:
definition of, 120, 199
Antibodies:
definition of, 163, 199
in milk, 163
Antioxidant:
butylated hydroxyanisole (BHA), 191
butylated hydroxytoluene (BHT), 191
definition of, 41, 199
sources of, 41

Anus:
definition of, 83, 200
location of, 82, 83
Appert, Nicolas, 189
Artificial:
definition of, 107, 200
Aspartame:
definition of, 107, 200
RDA, 108
sweetness of, 109–111
Atoms:
definition of, 17, 200
electrons of, 98, 205
neutral, 98, 212
protons of, 97, 215
Attract:
definition of, 98, 200

Bacteria:
affect on foods, 179
Clostridium botulinum, 181
Clostridium perfringens, 182
definition of, 10, 200
Escherichia coli, 182
fermentation by, 165
hemorrhagic *E. coli,* 182
illness caused by, 181–183
Lactobacillus bulgariucus, 171
in the mouth, 107
Salmonella, 182
Staphylococcus aureus, 181
Streptococcus thermophilus, 171
Baking powder:
definition of, 131, 200
double-acting, 132
leavening action, 131, 132, 134–137
Baking soda:
definition of, 131, 200

221